Non-Doubling Ahlfors Measures, Perimeter Measures, and the Characterization of the Trace Spaces of Sobolev Functions in Carnot-Carathéodory Spaces

Memoirs
of the American Mathematical Society

Number 857

Non-Doubling Ahlfors Measures,
Perimeter Measures, and the
Characterization of the Trace Spaces
of Sobolev Functions
in Carnot-Carathéodory Spaces

Donatella Danielli
Nicola Garofalo
Duy-Minh Nhieu

July 2006 • Volume 182 • Number 857 (first of 4 numbers) • ISSN 0065-9266

American Mathematical Society
Providence, Rhode Island

2000 *Mathematics Subject Classification.* Primary 43A85, 46E35; Secondary 35H20.

Library of Congress Cataloging-in-Publication Data

Danielli, Donatella, 1966–

Non-doubling Ahlfors measures, perimeter measures, and the characterization of the trace spaces of Sobolev functions in Carnot-Carathéodory spaces / Donatella Danielli, Nicola Garofalo, Duy-Minh Nhieu.

p. cm. — (Memoirs of the American Mathematical Society, ISSN 0065-9266 ; no. 857)

"Volume 182, number 857 (first of 4 numbers)."

Includes bibliographical references.

ISBN 0-8218-3911-X (alk. paper)

1. Sobolev spaces. 2. Measure theory. 3. Generalized spaces. I. Garofalo, Nicola. II. Nhieu, Duy-Minh, 1966– III. Title. IV. Series.

QA3.A57 no. 857

[QA+]

2006042821

Memoirs of the American Mathematical Society

This journal is devoted entirely to research in pure and applied mathematics.

Subscription information. The 2006 subscription begins with volume 179 and consists of six mailings, each containing one or more numbers. Subscription prices for 2006 are US$624 list, US$499 institutional member. A late charge of 10% of the subscription price will be imposed on orders received from nonmembers after January 1 of the subscription year. Subscribers outside the United States and India must pay a postage surcharge of US$31; subscribers in India must pay a postage surcharge of US$43. Expedited delivery to destinations in North America US$35; elsewhere US$130. Each number may be ordered separately; *please specify number* when ordering an individual number. For prices and titles of recently released numbers, see the New Publications sections of the *Notices of the American Mathematical Society*.

Back number information. For back issues see the *AMS Catalog of Publications*.

Subscriptions and orders should be addressed to the American Mathematical Society, P. O. Box 845904, Boston, MA 02284-5904, USA. *All orders must be accompanied by payment.* Other correspondence should be addressed to 201 Charles Street, Providence, RI 02904-2294, USA.

Copying and reprinting. Individual readers of this publication, and nonprofit libraries acting for them, are permitted to make fair use of the material, such as to copy a chapter for use in teaching or research. Permission is granted to quote brief passages from this publication in reviews, provided the customary acknowledgment of the source is given.

Republication, systematic copying, or multiple reproduction of any material in this publication is permitted only under license from the American Mathematical Society. Requests for such permission should be addressed to the Acquisitions Department, American Mathematical Society, 201 Charles Street, Providence, Rhode Island 02904-2294, USA. Requests can also be made by e-mail to `reprint-permission@ams.org`.

Memoirs of the American Mathematical Society is published bimonthly (each volume consisting usually of more than one number) by the American Mathematical Society at 201 Charles Street, Providence, RI 02904-2294, USA. Periodicals postage paid at Providence, RI. Postmaster: Send address changes to Memoirs, American Mathematical Society, 201 Charles Street, Providence, RI 02904-2294, USA.

© 2006 by the American Mathematical Society. All rights reserved.
Copyright of this publication reverts to the public domain 28 years
after publication. Contact the AMS for copyright status.
This publication is indexed in *Science Citation Index*®, *SciSearch*®, *Research Alert*®,
CompuMath Citation Index®, *Current Contents*®/*Physical, Chemical & Earth Sciences*.
Printed in the United States of America.

∞ The paper used in this book is acid-free and falls within the guidelines
established to ensure permanence and durability.
Visit the AMS home page at `http://www.ams.org/`

10 9 8 7 6 5 4 3 2 1   11 10 09 08 07 06

Dedicated to David Adams, on his 60th birthday

Contents

Chapter 1. Introduction	1
1.1. **Carnot-Carathéodory spaces**	9
1.2. The Chow-Rashevsky's accessibility theorem and CC metrics	9
1.3. The Nagel-Stein-Wainger polynomial and the size of the CC balls	11
Chapter 2. Carnot groups	15
2.1. Carnot groups of step 2	17
2.2. The Kaplan mapping	19
2.3. Groups of Heisenberg type	20
Chapter 3. The characteristic set	23
3.1. A result of Derridj on the size of the characteristic set	23
3.2. Some geometric examples	24
3.3. Non-characteristic manifolds	24
3.4. Manifolds with controlled characteristic set	27
Chapter 4. X-variation, X-perimeter and surface measure	33
4.1. The structure of functions in $BV_{X,loc}$	33
4.2. X-Caccioppoli sets	34
4.3. X-perimeter and the perimeter measure	36
Chapter 5. Geometric estimates from above on CC balls for the perimeter measure	37
5.1. A fundamental estimate	37
5.2. The X-perimeter of a $C^{1,1}$ domain is an upper 1-Ahlfors measure	39
Chapter 6. Geometric estimates from below on CC balls for the perimeter measure	41
6.1. The relative isoperimetric inequality and Theorem 6.1	42
6.2. A basic geometric lemma	43
6.3. Further analysis for Hörmander vector fields of step 2	46
6.4. Second proof of Theorem 6.1	52
6.5. Failure of the 1-Ahlfors condition for the X-perimeter of $C^{1,\alpha}$ domains	55
Chapter 7. Fine differentiability properties of Sobolev functions	57
7.1. Poincaré inequality, fractional integrals and improved representation formulas	57
7.2. Fine mapping properties of fractional integration on metric spaces	61
7.3. Differentiation with respect to an upper Ahlfors measure	62
7.4. Upper Ahlfors measures and Hausdorff measure	63

Chapter 8. **Embedding a Sobolev space into a Besov space with respect to an upper Ahlfors measure**   65
  8.1. Some results from harmonic analysis   65
  8.2. Two simple growth-estimates   66
  8.3. A key continuity estimate for a singular integral   67
  8.4. The main theorem   69

Chapter 9. **The extension theorem for a Besov space with respect to a lower Ahlfors measure**   79
  9.1. Some auxiliary results   79
  9.2. Proof of Theorem 9.1   80

Chapter 10. **Traces on the boundary of (ϵ, δ) domains**   85
  10.1. The (ϵ, δ) condition is optimal for the existence of traces   88
  10.2. Characterization of the traces on the boundary   90

Chapter 11. **The embedding of $B^p_\beta(\Omega, d\mu)$ into $L^q(\Omega, d\mu)$**   93

Chapter 12. **Returning to Carnot groups**   99

Chapter 13. **The Neumann problem**   103

Chapter 14. **The case of Lipschitz vector fields**   109

Bibliography   111

Abstract

The object of the present study is to characterize the traces of the Sobolev functions in a sub-Riemannian, or Carnot-Carathéodory space. Such traces are defined in terms of suitable Besov spaces with respect to a measure which is concentrated on a lower dimensional manifold, and which satisfies an Ahlfors type condition with respect to the standard Lebesgue measure. We also study the extension problem for the relevant Besov spaces. Various concrete applications to the setting of Carnot groups are analyzed in detail and an application to the solvability of the subelliptic Neumann problem is presented.

Received by the editor February 23, 2002.

1991 *Mathematics Subject Classification*. Primary 43A85, 46E35; Secondary 35H20.

Key words and phrases. traces, restriction, extension, Sub-elliptic Sobolev spaces, Besov spaces, perimeter measures.

First author supported in part by NSF grant DMS-0002801 and by NSF CAREER Award DMS-0239771.

Second author supported in part by NSF Grant No. DMS-0070492 and by NSF Grant No. DMS-0300477.

Third author supported in part by NSC Grant No. 89-2115-M-001-018.

CHAPTER 1

Introduction

In the last decade there has been an explosion of interest in the theory of *Carnot-Carathéodory spaces* (CC spaces, henceforth), and in the ramifications of this subject into analysis and geometry. We recall that a CC space is a Riemannian manifold (M, g), which has been endowed with a distance d different from the Riemannian metric attached to the tensor g. Such distance d is the *control metric* associated with a sub-bundle H of the tangent bundle TM. Loosely speaking, if $X = \{X_1, ..., X_m\}$ denotes a system of non-commuting vector fields which (locally) generates H, then one defines $d(x, y)$ by a minimization procedure which selects among all curves in M which join x to y, only those whose tangent vector belongs to $\text{span}\{X_1, ..., X_m\}$. The ensuing metric space (M, d) is called a CC space, or also a *sub-Riemannian space*. Excellent references on the subject are the books [**VSC92**], [**Be96**], [**Gro98**], [**Mon02**]. We also refer the reader to the forthcoming book [**G02**], which more directly treats the connections of CC geometry with partial differential equations. In this work we study the following general question: to characterize the traces of Sobolev functions in a CC space with respect to a measure supported on a lower dimensional manifold. Our primary motivation is the study of boundary value problems, arising for instance in CR geometry, for various linear and nonlinear equations of sub-elliptic type. The common trend of these equations is lack of ellipticity. But their leading part can be expressed by the sum of squares of smooth vector fields satisfying a certain algebraic assumption known as the finite rank condition on the Lie algebra, see (1.4). Thanks to a fundamental result of Hörmander [**H67**], such condition implies the hypoellipticity of the relevant operator, see also [**OR73**].

Another important motivation is the connection of the questions studied here with the newly forming theory of perimeters and minimal surfaces in CC spaces. We recall that the existence of minimal surfaces was established in [**GN96**], where the problem of their regularity was also posed. This aspect, however, is only barely touched upon in the present work, and will be systematically investigated in forthcoming studies. In this connection we mention the interesting recent papers [**FSS01**], [**Pa(II)04**], [**FSS03(I)**], [**FSS03(II)**], and also the preprint [**DGN04(II)**].

Among many others devoted to the subject, the classical books [**Ne67**], [**LaU68**], [**LiMa72**], [**Tre75**], [**Gr85**], [**Tro87**], underline the fundamental role played by trace theorems in the theory of boundary value problems for partial differential equations. The reader should also consult the pioneering papers by Gagliardo [**Ga57**], [**Ga58**], [**Ga59**], and Stein [**St61**]. The more recent papers [**JK95**], [**FMM98**] and [**MiMa04**] contain sharp results for the solvability of boundary value problems for the non-homogeneous Laplace equation in the framework of Besov spaces.

To introduce the problems studied in the present work consider the Euclidean space $\mathbb{R}^n$ with Lebesgue measure dx. If $\Omega \subset \mathbb{R}^n$ denotes a Lipschitz domain, and if we indicate with $d\mu = dH_{n-1}\lfloor \partial\Omega$, the restriction of the ordinary $(n-1)$-dimensional Hausdorff measure to the boundary of Ω, then it is well known that there exist constants $0 < \alpha < \beta$, depending only on n, and on the Lipschitz character of Ω, such that for every $x_o \in \partial\Omega$, and any $r > 0$, one has
$$\alpha \, r^{n-1} \leq \mu(B_e(x_o, r)) = \mu(\partial\Omega \cap B_e(x_o, r)) \leq \beta \, r^{n-1} \, ,$$
where we have let $B_e(x_o, r) = \{x \in \mathbb{R}^n \mid |x - x_o| < r\}$. This property can be more suggestively reformulated as follows

$$(1.1) \qquad \alpha' \, \frac{|B_e(x_o, r)|}{r} \leq \mu(B_e(x_o, r)) \leq \beta' \, \frac{|B_e(x_o, r)|}{r} \, ,$$

where from now on $|E|$ indicates the n-dimensional Lebesgue measure of E. The inequality (1.1) represents an important property of Lipschitz domains. It has several deep repercussions, among which the well-known trace inclusion for the classical Sobolev spaces

$$(1.2) \qquad W^{1,p}(\Omega) \subset L^q(\partial\Omega, d\mu) \, , \qquad q = p(n-1)/(n-p),$$

valid when $1 \leq p < n$. We stress that the exponent q is sharp in the scale of Lebesgue spaces. Moreover, one has $q > p$ when $1 < p < n$, whereas there is no gain at the end-point $p = 1$, since in that case one has $q = 1$ as well. The embedding (1.2) says that, despite the fact that the boundary of Ω is a set of minimal smoothness, and of Lebesgue measure zero, it is nonetheless possible to define the *trace* of a Sobolev function on $\partial\Omega$. Moreover, the latter belongs to a Lebesgue space with respect to the measure μ. As it turns out, for this result one does not need the full strength of (1.1), but only the estimate from above. This was discovered in a beautiful paper of D. Adams [**A71**], see also [**A73**], [**AH96**], who brought a new perspective into the problem. For the first time, it became apparent that one of the central elements of the classical embedding theorems are size estimates - such as (1.1) - of the measure μ in the target space. In this perspective, the problem of traces is divided into two main steps: 1. Establishing the embedding under such *a priori* imposed size estimates; 2. Finding *good* geometric conditions on the support of the measure μ which guarantee the validity of such estimates. Clearly, in the step 2 the case in which the measure μ is the surface measure on the boundary of the ground domain is of great interest.

Such ideas have had a lasting influence, and we will see them resurface in the present work. The case $p = 1$ of (1.2) is connected with geometric measure theory and plays a key role in the study of minimal surfaces. Here, the appropriate substitute for the Sobolev space $W^{1,1}(\Omega)$ is the space $BV(\Omega)$ of functions having bounded variation in Ω, see [**Gi84**], [**Zi89**], [**EG92**]. The existence of traces of BV functions was elegantly settled in [**MZ77**], which was also influenced by the above described approach in Adams' paper [**A71**]. Again, the upper estimate in (1.1) plays an essential role. However, the proof of the main trace inequality is more delicate than its Sobolev space counterpart since it also crucially relies on the structure of sets with finite perimeter. For this latter aspect we refer the reader to the basic results in [**FSS01**], [**FSS03(I)**].

Once the existence of traces is ascertained, it is natural to ask whether they themselves possess any degree of smoothness, and if so, whether it is possible to give a complete characterization of them. It is well-known that the answer to these questions is provided by the fractional Sobolev, or Besov, space $W^{1-\frac{1}{p},p}(\partial\Omega, d\mu)$. A function $f \in L^p(\partial\Omega, d\mu)$ is said to belong to such space if the semi-norm

$$(1.3) \qquad \left\{\int_{\partial\Omega}\int_{\partial\Omega}\left(\frac{|f(x)-f(y)|}{|x-y|^\beta}\right)^p \frac{1}{|x-y|^{n-1}} d\mu(y)\, d\mu(x)\right\}^{\frac{1}{p}}$$

is finite. Here, $\beta = 1 - 1/p$ is the fractional order of differentiation. It is a classical fact that in order to characterize $W^{1-\frac{1}{p},p}(\partial\Omega, d\mu)$ as the trace space of $W^{1,p}(\Omega)$, one also needs to construct an extension operator. To accomplish this task, the estimate from below in (1.1) becomes important, as well as several other tools from harmonic analysis. An excellent reference for these aspects is the monograph [**JW84**]. Our goal is to generalize the above mentioned classical results to the Sobolev spaces $\mathcal{L}^{1,p}(\Omega)$ associated with a system $X = \{X_1, ..., X_m\}$ of vector fields in $\mathbb{R}^n$. As we will see, this endeavor entails analyzing in depth several new problems that, in the classical Euclidean setting, do not appear, or that are easily resolved. In the process, new tools must be developed.

To fix the ideas, let us consider a system $X = \{X_1, ..., X_m\}$ of C^∞ vector fields in $\mathbb{R}^n$, which we assume equipped with the standard Lebesgue measure dx. If the finite rank condition on the Lie algebra is fulfilled,

$$(1.4) \qquad rank\ Lie[X_1, ..., X_m] \equiv n\ ,$$

than thanks to the accessibility Theorem 1.4 of Chow-Rashevsky [**Ch39**], [**Ra38**], the CC metric associated with the system X is well-defined. A basic theorem of Nagel-Stein-Wainger [**NSW84**] states that such distance is locally uniformly *doubling* with respect to Lebesgue measure, see Theorem 1.12. We consider next the collection of all Borel measures $\mathcal{B}_d$ on the metric space $(\mathbb{R}^n, d)$.

Problem: *For which $\mu \in \mathcal{B}_d$, and exponents $1 \leq p \leq q < \infty$, does the a priori inequality*

$$(1.5) \qquad \left(\int_B |u - u_{B,\mu}|^q\, d\mu\right)^{\frac{1}{q}} \leq C \left(\int_{B^*} |Xu|^p\, dx\right)^{\frac{1}{p}}, \qquad u \in C^\infty(B^*),$$

hold?

The notation $B = B(x_o, r)$ in (1.5) indicates the ball centered at x_o with radius $r > 0$ in the CC metric d, whereas $B^* = B(x_o, \sigma R)$, $\sigma \geq 1$. The symbol $u_{B,\mu}$ stands for the μ-average of u over B. Finally, we have denoted with $Xu = (X_1 u, ..., X_m u)$ the *sub-gradient* of u along the system X, so that $|Xu| = (\sum_{j=1}^m (X_j u)^2)^{1/2}$. The above problem, which constitutes the sub-Riemannian analog of (1.2), is important in the study of boundary value problems for sub-elliptic equations arising in several complex variables, in CR geometry (e.g., in the study of the CR Yamabe problem), in the study of quasi-conformal mappings between nilpotent Lie groups, in control theory, and last, but not least, in the development of geometric measure theory in CC spaces, particularly, in the theory of minimal surfaces. For the case in which the measure μ admits a density with respect to Lebesgue measure, i.e., $d\mu = V dx$, a sharp trace inequality was proved in [**D99**] under the hypothesis that the density V belongs to a suitable Morrey-Campanato space with respect to the CC distance. Subsequently, in the paper [**DGN98**] we were able to provide a complete answer to the problem stated above. We distinguished between the *geometric case*, corresponding to $p = 1$, and the *non-geometric case*, when $p > 1$. Concerning the case $p > 1$, the main result in [**DGN98**] was the following.

THEOREM 1.1. *Consider a bounded, open set $U \subset \mathbb{R}^n$, with local homogeneous dimension Q. For $1 < p < Q$, let μ be a nonnegative Borel measure on $\mathbb{R}^n$ such that for some $M > 0$, $0 \leq s < p$, and $R_o > 0$, one has*

$$(1.6) \qquad \mu(B(x,r)) \leq M \frac{|B(x,r)|}{r^s}, \qquad x \in U, \ 0 < r \leq R_o.$$

There exist positive constants, $C = C(U, X, p, s)$ and $\sigma = \sigma(U, X) \geq 1$, such that for any $x_o \in U$, $0 < R \leq R_o$, $B = B(x_o, R)$, $\sigma B = B(x_o, \sigma R)$, the following holds: If $u \in \mathcal{L}^{1,p}(\sigma B)$, then there exists a uniquely determined $\tilde{u} \in L^q(B, d\mu)$, where $q = p\frac{Q-s}{Q-p} > p$, such that

$$\left(\int_B |\tilde{u} - \tilde{u}_{B,\mu}|^q \, d\mu\right)^{\frac{1}{q}} \leq C\, M^{\frac{1}{q}} \left(\frac{R}{|B|^{\frac{1}{Q}}}\right)^{\frac{(p-s)Q}{p(Q-s)}} \left(\int_{\sigma B} |Xu|^p \, dx\right)^{\frac{1}{p}}.$$

The function $\tilde{u}$ is called the trace of u in $L^q(B, d\mu)$. *Finally, (1.6) is also necessary for the latter inequality to hold.*

We remark that Theorem 1.1 incorporates the optimal Sobolev embeddings in [**D92**], [**Lu94**], [**MaSC95**], [**BM95**]. To see this it is enough to take $d\mu = dx$, and notice that (1.6) trivially holds with $M = 1$ and $s = 0$. One thus concludes, with obvious meaning of the notations,

$$\left(\frac{1}{|B|} \int_B |u - u_B|^q \, dx\right)^{\frac{1}{q}} \leq C\, R \left(\frac{1}{|B|} \int_{\sigma B} |Xu|^p \, dx\right)^{\frac{1}{p}},$$

where now $q = pQ/(Q-p)$. We note explicitly that in the statement of Theorem 1.1 no control from below is imposed on the measure μ. In particular, we *do not* assume that μ be a *doubling measure*. In the case $p = 1$, using ideas from geometric measure theory, we established a corresponding end-point result, see Theorem 1.4 in [**DGN98**].

Although these results are sharp in the scale of the Lebesgue spaces $L^q(B, d\mu)$, it is of paramount importance to be able to identify the precise trace space on the boundary for a Sobolev function. In this paper we introduce an appropriate class of Besov spaces with respect to a given distance d and a given $\mu \in \mathcal{B}_d$, and we prove that, under some natural growth assumptions on μ modeled on (1.1), *these are the trace spaces* for the sub-elliptic Sobolev spaces $\mathcal{L}^{1,p}(\Omega)$. We also analyze in depth the basic problem of the examples. We will not discuss here, however, the trace space of BV functions introduced in [**GN96**].

The following definition plays a pervasive role in the sequel.

DEFINITION 1.2. *Given $s \geq 0$, a measure $\mu \in \mathcal{B}_d$ will be called an* upper s-Ahlfors measure, *if there exist $M, R_o > 0$, such that for $x \in \mathbb{R}^n$, $0 < r \leq R_o$, one has*

$$(1.7) \qquad \mu(B(x,r)) \leq M \frac{|B(x,r)|}{r^s}.$$

We will say that μ is a *lower s-Ahlfors measure*, if for some $M, R_o > 0$ one has instead for x and r as above

$$(1.8) \qquad \mu(B(x,r)) \geq M^{-1} \frac{|B(x,r)|}{r^s}.$$

When μ is both an upper and lower s-Ahlfors measure, then we say that it is a s-Ahlfors measure.

We thus come to the central definition for the results in this paper.

DEFINITION 1.3. Given a distance d, let $\mu \in \mathcal{B}_d$ be an upper (or lower) s-Ahlfors measure, having $supp\ \mu \subseteq F$, where F is a closed subset of $\mathbb{R}^n$. For $1 \leq p < \infty$, $0 < \beta < 1$, we introduce the semi-norm

$$\mathcal{N}_\beta^p(f, F, d\mu) = \left\{ \int_F \int_F \left(\frac{|f(x) - f(y)|}{d(x,y)^\beta} \right)^p \frac{d(x,y)^s}{|B(x, d(x,y))|} d\mu(y)\, d\mu(x) \right\}^{\frac{1}{p}}.$$

The *Besov space* on F, relative to the measure μ, is defined as

$$B_\beta^p(F, d\mu) = \{ f \in L^p(F, d\mu) \mid \mathcal{N}_\beta^p(f, F, d\mu) < \infty \}.$$

If $f \in B_\beta^p(F, d\mu)$, we define the Besov norm of f as

$$\|f\|_{B_\beta^p(F, d\mu)} = \|f\|_{L^p(F, d\mu)} + \mathcal{N}_\beta^p(f, F, d\mu).$$

To motivate Definition 1.3 we observe that, when $X = \{\partial/\partial x_1, ..., \partial/\partial x_n\}$ is the standard basis of $\mathbb{R}^n$, then one easily sees that the ensuing CC metric is just the ordinary Euclidean distance $d_e(x, y) = |x - y|$. If $\Omega \subset \mathbb{R}^n$ is a bounded Lipschitz domain, and we take $\mu = H_{n-1} \lfloor \partial \Omega$, then (1.1) shows that μ is a 1-Ahlfors measure with respect to d_e. If we thus let $F = \partial \Omega$, then with $s = 1$ the semi-norm $\mathcal{N}_\beta^p(f, F, d\mu)$ gives back the classical Besov semi-norm in (1.3). We emphasize the purely metrical character of the semi-norm $\mathcal{N}_\beta^p(f, F, d\mu)$. The reader will have noticed that the parameter s, which is present in the right-hand side of the definition of the semi-norm, yet does not appear in the left-hand side $\mathcal{N}_\beta^p(f, F, d\mu)$. This omission is intentional, and is dictated by the desire of not overburdening the notation. It will not lead to any confusion since the specific assumptions on μ and s will be clearly spelled in the statement of each theorem. Finally, we note that the parameter β in Definition 1.3 measures "smoothness". In this respect, it is interesting to observe that if (1.7) holds, and if $\beta' > \beta$, then the sub-elliptic Hölder class $\Gamma^{0,\beta'}(F)$ (for whose definition in the context of homogeneous groups we refer to [**FS74**], [**F75**]) is continuously embedded into $B_\beta^p(F, d\mu)$.

A description of the present work can be found in the table of contents. The reader is also referred to the individual chapters for a discussion of the bibliographical accounts. We add here a few clarifying comments. Chapter 1.1 contains various preliminary notions and basic results about CC distances. The main body of the paper starts with Chapter 2, where we discuss those infinitesimal Lie groups which constitute the fundamental models of CC spaces. From that point on, the treatment is essentially divided into two parts. The former, Chapters 2-6, is preparatory to the latter, Chapters 7-12, and also has an independent interest. In Chapters 2-6 we focus on constructing geometric examples of upper/lower Ahlfors measures. Keeping in mind that the question of examples is of paramount importance, it should not be surprising that we dedicate to it considerable effort. In CC geometry a crucial notion is that of *characteristic point* on the boundary of a given domain, see Chapter 3. Typically, bounded domains do have characteristic points. For instance, in the Heisenberg group $\mathbb{H}^n$ every bounded C^1 domain which is homeomorphic to the unit sphere $\mathbb{S}^{2n} \subset \mathbb{R}^{2n+1}$ must have non-empty characteristic set. At characteristic points the vector fields $X_1, ..., X_m$ become tangent to the boundary and most of the tools from classical analysis fail to work. For example, near the characteristic set standard surface measure does not scale correctly and fails to satisfy size estimates such as (1.1) with respect to the CC balls. In [**DGN98**] we proved that the *ad hoc* replacement of surface measure is the X-perimeter $P_X(E;\cdot)$, introduced in [**CDG94**]. The latter generalizes the notion of perimeter according to De Giorgi, see Chapter 4, and appears naturally in the intrinsic relative isoperimetric inequalities on CC balls established in [**GN96**], as well as in the first and second variation formula for minimal surfaces, see [**DGN04(II)**]. An essential feature of the X-perimeter is that, unlike surface measure, it incorporates the geometric properties of the boundary near its characteristic set. To explain this point we notice that when E is a C^1 domain with outward unit normal ν, then it was proved in [**CDG94**] that for any open set Ω one has

$$(1.9) \qquad P_X(E;\Omega) = \int_{\partial E \cap \Omega} |\vec{X_\nu}| \, dH_{n-1} ,$$

where we have let $\vec{X_\nu} = (<X_1,\nu>, ..., <X_m,\nu>)$. Since at a characteristic point one has $|\vec{X_\nu}| = 0$, it is reasonable to expect that the X-perimeter should be the appropriate measure on the boundary. Proving this intuition correct requires a great deal of work.

Concerning upper Ahlfors measures the principal results are Theorems 5.3, 5.5, whose main consequence, Theorem 5.6, can be summarized by saying that, given a $C^{1,1}$ domain $\Omega \subset \mathbb{R}^n$ of type ≤ 2, the X-perimeter measure $P_X(\Omega;\cdot)$ is an upper 1-Ahlfors measure with respect to the control metric d associated with X. A C^1 domain $\Omega = \{x \in \mathbb{R}^n \mid \phi < 0\}$ is called of *type* ≤ 2 if for every characteristic point $x_o \in \partial\Omega$ there exist $i,j = 1,...,m$ such that $[X_i,X_j]\phi(x_o) \neq 0$, see also Definition 5.1. Clearly, if Ω has empty characteristic set, then for every boundary point x_o there exists $i \in \{1,...,m\}$ such that $X_i\phi(x_o) \neq 0$, and therefore Ω is of type 1. We stress that in a Carnot group of step $r = 2$ every C^1 domain is automatically of type ≤ 2, see Lemma 6.7, and therefore the type assumption imposes in this setting no restrictions on the characteristic set. In this perspective, we recall that for the Heisenberg group $\mathbb{H}^n$, Theorems 5.3 and 5.6 were first proved in [**DGN98**]. They were subsequently extended to Carnot groups of step 2 in [**CGN02**]. In both papers the type condition was not present, since, as we have explained, the latter becomes relevant only for groups of step $r \geq 3$. The general results in Theorems 5.3, 5.5 have been obtained in recent joint work of the second named author with L. Capogna [**CG05**]. In the same paper it is also proved that the type assumption in Theorems 5.3, 5.5 and 5.6 is best possible. In fact, in [**CG05**] the authors give an example of a C^∞ domain of type 3 in a group of step 3 for which the upper estimates in Theorems 5.3 and 5.6 fail, and consequently the X-perimeter fails to be an upper 1-Ahlfors measure.

The estimate from below of the perimeter measure has been for a long time an intriguing open question. Keeping (1.9) in mind, and recalling that at characteristic points $|\vec{X_\nu}|$ vanishes, it should be clear to the reader that estimating $P_X(\Omega;\cdot)$ from below is a very delicate task. In this paper we give a contribution to this problem. Our main result is Theorem 6.1, which states that in a $C^{1,1}$ domain in a Carnot group of step 2 the X-perimeter is a lower 1-Ahlfors measure. Combining this result with Theorem 5.6 we conclude that for any $C^{1,1}$ domain in a Carnot group of step 2, the X-perimeter is a 1-Ahlfors measure. In particular, $P_X(\Omega;\cdot)$ is doubling. Again, no restriction on the characteristic set is present in this result. We emphasize that the $C^{1,1}$ smoothness of the domain in Theorem 6.1 is best possible. As we show in the sub-chapter 6.5, for every $0 < \alpha < 1$ there exists a $C^{1,\alpha}$ domain in the Heisenberg group whose perimeter measure fails to be lower 1-Ahlfors. In this negative phenomenon, what is lurking in the dark is the delicate balance between characteristic points and the non-isotropic group dilations.

The second part of this work, Chapters 7-12, is devoted to establishing the various trace and extension theorems connecting the Sobolev spaces $\mathcal{L}^{1,p}(\Omega, dx)$ to the Besov spaces $B^p_\beta(F, d\mu)$. In Chapter 7 we address a basic question which is, in a sense, preliminary to the study of traces. In analyzing the existence of traces we work with an upper Ahlfors measure μ. Since the latter is generally supported on a set of zero Lebesgue measure dx, there is no guarantee that a funtion in a Sobolev space with respect to dx be defined on the support of μ. In Theorem 7.8, which constitutes a refinement of Lebesgue differentiation theorem, we prove that this is actually the case, thus putting the study of the trace problem on a firm ground. The latter begins in Chapter 8, where we prove that, given an upper s-Ahlfors measure μ, with $0 < s < p$, a function in $\mathcal{L}^{1,p}$ with respect to dx possesses a trace in the optimal Besov space $B^p_{1-\frac{s}{p}}$ with respect to μ. The main result of the chapter is Theorem 8.6. The proof of this theorem involves a substantial amount of work, and combines various ideas from harmonic analysis. Theorem 8.6 is local in nature, in the sense that it establishes the existence of traces when the upper Ahlfors measure μ is supported inside the domain of the relevant functions. Subsequently in the chapter, see Theorem 8.8, we consider the case in which μ is the X-perimeter measure on an interior boundary. Chapter 9 is devoted to the delicate task of constructing an extension operator from a Besov to a Lebesgue space. Our main result in this direction is Theorem 9.1. Contrarily to the trace Theorem 8.6, here the key assumption is that μ be a lower Ahlfors measure. In Chapter 10 we characterize the traces on the boundary for the general class of (ϵ, δ) domains with respect to a metric d. When $d(x,y) = |x-y|$, such notion coincides with that of *uniform domain* introduced by Martio and Sarvas [**MaSa78**]. In [**Jo81**] P. Jones established extension theorems for the ordinary Sobolev spaces on (ϵ, δ) domains. In Theorem 10.6 we prove that, given $p > 1$, there exists a continuous trace operator

(1.10) $$\mathcal{T}r \ : \ \mathcal{L}^{1,p}(\Omega, dx) \ \to \ B^p_{1-\frac{s}{p}}(\partial\Omega, d\mu) \ ,$$

when μ is an upper s-Ahlfors measure, for some $0 < s < p$. When $\mu = P_X(\Omega;\cdot)$, we can combine Theorem 10.6 with Theorem 5.6, and establish (1.10) with $s = 1$, see Theorem 10.7. The crowning result of the chapter is Theorem 10.9, in which we finally characterize $B^p_{1-\frac{1}{p}}(\partial\Omega, d\mu)$ as *the trace space* of the Sobolev space $\mathcal{L}^{1,p}(\Omega, dx)$. For this result, one needs to assume that μ is a s-Ahlfors measure. Concerning the (ϵ, δ) assumption in Theorem 10.6, we prove in Proposition 10.8 that it cannot be weakened. In Chapter 11 we close the gap between the above cited Theorem 1.1 from [**DGN98**], and Theorem 8.6. We prove that, when μ is a lower s-Ahlfors measure supported in $F \subset \Omega$, then the space $B^p_\beta(F, d\mu)$ is continuosly embedded into an optimal Lebesgue space $L^q(\Omega, d\mu)$. Combining this result with (1.10) in Theorem 10.6, we recover Theorem 1.1, except that we have made the additional assumption on μ of being lower s-Ahlfors. Finally, in Chapter 12 we apply the theory developed in parts one and two to the setting of Carnot groups. Using the perimeter measure on the boundary of "minimally smooth" domains, we obtain some interesting concrete examples of the main results in this work. Chapters 13 and 14 conclude this paper. In the former we give an application of our trace theorem by establishing the existence and uniqueness (modulo constants) of the variational solution of the Neumann problem for sub-Laplacians. We stress that this is not just a functional analytic theorem, since it heavily relies on both the results of part one and two discussed above. The study of the Neumann problem will be taken-up in a future study. The purpose of the short, conclusive Chapter 14 is to bring to the reader's attention that, with one natural additional assumption, our results continue to hold in the more general setting of Lipschitz vector fields analyzed in [**DGN98**]. The reader will have no difficulty in realizing that, given the general nature of our approach, the results in Chapters 7-12 carry over to the setting of metric spaces with a doubling measure and a Poincaré inequality. In this context, the appropriate notion of gradient can be given using that introduced by Hajlasz, or that due to Heinonen and Koskela. We do not explicitly treat these aspects, but refer the reader to [**HK98**], [**Gro98**], and also to the papers [**Ha96**], [**HaM97**], [**Che99**].

In closing, we mention some references which are connected to the present work. In his Ph.D. Dissertation M. Mekias [**Me93**] studied the problem of traces in the Heisenberg group $\mathbb{H}^n$. Although specialized to this setting, his work contains several sharp results which are closely connected to some of ours. Mekias also recognized the relevance of the measure μ introduced in Definition 4.7, and established several of its properties, although he was not aware of the fact that such μ expresses the X-perimeter measure. We also mention the Ph.D. Dissertation of C. Romero [**Ro91**] which contains an earlier version of Theorem 5.3 for the Heisenberg group. In the paper [**BP99**], Berhanu and Pesenson established some trace and extension theorems for C^∞ vector fields satisfying the finite rank condition (1.4) at step 2. Their work, however, only treats non-characteristic manifolds, with two additional hypothesis on the vector fields. Also, their definition of Besov semi-norm is built on these assumptions, and differs from our Definition 1.3. Again for non-characteristic domains, and for the case of step 2 vector fields, Bahouri, Chemin and Xu [**BCX99**] characterized the traces in L^2 using techniques from microlocal analysis (Weyl-Hörmander calculus). Their definition of Besov space is quite different from ours, since it is given interpolating between standard Sobolev spaces. For the special setting of the Heisenberg group $\mathbb{H}^n$ they are also able to establish a trace theorem when the manifold possesses only isolated characteristic points. Finally, in their recent paper [**MM02**] Monti and Morbidelli have proved the embedding (1.10), but not the characterization of the traces, when $\Omega \subset \mathbb{R}^n$ is a bounded C^∞ domain with non-characteristic boundary. They use the Besov semi-norm introduced in Definition 1.3 (in fact, a modification of the latter), except that they work with the ordinary surface measure $\mu = H_{n-1} \lfloor \partial\Omega$, instead of the perimeter measure. We notice that when $\partial\Omega$ is non-characteristic, this μ is obviously equivalent to the measure $P_X(\Omega; \cdot)$ in (1.9). It is interesting to compare their result with our Theorem 10.7. We do not require that $\partial\Omega$ be non-characteristic, and furthermore we only assume minimal $C^{1,1}$ smoothness of Ω. This is possible thanks to Theorem 5.5, to the extension theorem for Sobolev spaces proved in [**GN98**], and to our Theorem 8.6 which we discussed above. To apply the extension theorem, the (ϵ, δ)

condition is needed. In this respect we emphasize that in [**MM04(II)**] Monti and Morbidelli have proved that in the general Hörmander case any C^∞, non-characteristic bounded domain is NTA with respect to the CC metric, hence in particular it is (ϵ, δ). Since every non-characteristic domain is trivially of type ≤ 2, as a consequence of these considerations, the main result in [**MM02**] becomes a special case of Theorem 10.7. On the other hand, it must be said that the assumption that $\partial \Omega$ be non-characteristic in [**MM02**] allows to avoid resorting to the extension procedure by working directly on Ω. Monti and Morbidelli also treat an example of characteristic domain for the special situation of the Baouendi-Grushin vector fields in the plane $X_1 = \partial/\partial x$, $X_2 = |x|^\alpha \partial/\partial y$, $\alpha > 0$. For a discussion of the latter we refer the reader to Chapter 3.

1.1. Carnot-Carathéodory spaces

In this chapter we collect some definitions and various basic known results which are used in the main body of the paper.

1.2. The Chow-Rashevsky's accessibility theorem and CC metrics

Let $X = \{X_1, ..., X_m\}$ be a system of C^∞ vector fields in $\mathbb{R}^n$, $n \geq 3$, satisfying the finite rank condition (1.4). A piecewise C^1 curve $\gamma : [0, T] \to \mathbb{R}^n$ is called *sub-unitary* if for every $t \in (0, T)$ for which $\gamma'(t)$ exists one has

$$(1.11) \qquad <\gamma'(t), \xi>^2 \leq \sum_{j=1}^n <X_j(\gamma(t)), \xi>^2 \qquad \text{for every} \quad \xi \in \mathbb{R}^n.$$

The reader should notice that definition (1.11) forces the condition

$$\gamma'(t) \in span \{X_1(\gamma(t)), ..., X_1(\gamma(t))\} .$$

We define the sub-unitary length of γ as $l_s(\gamma) = T$. Given $x, y \in \mathbb{R}^n$, denote by $\mathcal{S}_U(x, y)$ the collection of all sub-unitary $\gamma : [0, T] \to U$ which join x to y. We will need the following fundamental accessibility theorem due Chow [**Ch39**] and Rashevsky [**Ra38**].

THEOREM 1.4. *Given a connected open set $U \subset \mathbb{R}^n$, for every $x, y \in U$ there exists $\gamma \in \mathcal{S}_U(x, y)$.*

As a consequence of Theorem 1.4, if we pose

$$(1.12) \qquad d_U(x, y) = \inf \{l_s(\gamma) \mid \gamma \in \mathcal{S}_U(x, y)\},$$

we obtain a distance on U, called the *Carnot-Carathéodory distance on U* associated with the system X. When $U = \mathbb{R}^n$, we write $S(x, y)$, instead of $S_{\mathbb{R}^n}(x, y)$, and $d(x, y)$, instead of $d_{\mathbb{R}^n}(x, y)$. It is clear that

$$(1.13) \qquad d(x, y) \leq d_U(x, y) \qquad x, y \in U,$$

for every connected open set $U \subset \mathbb{R}^n$. For $x \in \mathbb{R}^n$, and $r > 0$, we let $B(x, r) = \{y \in \mathbb{R}^n \mid d(x, y) < r\}$. We indicate with $B_e(x, r) = \{y \in \mathbb{R}^n \mid |x - y| < r\}$ the corresponding Euclidean ball. When $x \in U$, we will write instead $B_U(x,r) = \{y \in U \mid d_U(x,y) < r\}$. The following elementary property of d_U will be useful, see, e.g., [**NSW84**].

PROPOSITION 1.5. *For every connected set $U \subset\subset \mathbb{R}^n$, there exists $C = C(U,X) > 0$ such that*
$$|x-y| \leq C\, d_U(x,y) \qquad \text{for every} \quad x,y \in U.$$
This gives for every $x \in U$, and any $r > 0$,
$$\tag{1.14} B_U(x,r) \subset B_e(x,Cr).$$

Another consequence of Theorem 1.4 is the following property noted in [**GN96**].

PROPOSITION 1.6. *One has*
$$i : (\mathbb{R}^n, d) \to (\mathbb{R}^n, |\cdot|) \qquad \text{is continuous.}$$

The next basic result, established in [**RS76**], see also [**NSW84**], provides a quantitative version of accessibility.

THEOREM 1.7. *Given a connected set $U \subset\subset \mathbb{R}^n$, there exist $C = C(U,X) > 0$, and $\epsilon = \epsilon(U,X) > 0$, such that for every $x,y \in U$*
$$d_U(x,y) \leq C^{-1}\, |x-y|^\epsilon.$$
This gives for every $x \in U$ and $r > 0$
$$B_e(x, (Cr)^{1/\epsilon}) \subset B_U(x,r).$$

A deep theorem of C. Fefferman and D. H. Phong [**FP81**] states that the inclusion between Euclidean and CC balls in the statement of Theorem 1.7 is, in fact, a necessary and sufficient condition for the validity of sub-elliptic estimates for a general class of operators with smooth coefficients and non-negative characteristic form. Theorem 1.7, coupled with (1.13), gives
$$\tag{1.15} d(x,y) \leq C^{-1}\, |x-y|^\epsilon \qquad x,y \in U\ ,$$
and from this we obtain the following important property.

PROPOSITION 1.8. *The inclusion*
$$i\ :\ (\mathbb{R}^n, |\cdot|) \to (\mathbb{R}^n, d)$$
is continuous .

As a consequence of Proposition 1.6, and of Proposition 1.8, the Euclidean and the CC topology on $\mathbb{R}^n$ coincide. In particular, compact sets with respect to either topology are the same. However, if the vector fields $\{X_j\}_{j=1,\ldots,m}$ grow at infinity faster than linearly, then the compactness of metric balls of large radii may fail in general, see [**GN96**], [**G02**]. Consider for instance in $\mathbb{R}$ the smooth vector field $X = (1+x^2)\, d/dx$. Elementary calculations prove that the CC distance relative to X is given by
$$d(x,y) = |\arctan x - \arctan y|,$$

and therefore, if $r \geq \pi/2$, we have $B(0,r) = \mathbb{R}$. This global aspect is intimately connected to the powerful extension of the Theorem of Hopf-Rinow due to Cohn-Vossen [**CV35**], and we refer the reader to the forthcoming book [**G02**] for a detailed discussion. In the present paper we are solely concerned with local questions. Thereby, in order to eliminate all the topological complications connected to the growth of the vector fields at infinity, we will henceforth make the following hypothesis:

(1.16) $\qquad$ *The vector fields $X_1,...,X_m$ have coefficients in $Lip(\mathbb{R}^n)$* .

Such assumption will be in force throughout the paper in the purely Hörmander case. It is instead unnecessary for Carnot groups since, in that framework, the compactness of balls holds irregardless of the radius, and of the growth of the X_j's at infinity, see [**G02**]. A basic consequence of (1.16) is the following result established in [**GN98**].

PROPOSITION 1.9. *Under the hypothesis (1.16), for any $x_o \in \mathbb{R}^n$, and every $r > 0$, the closed ball $\overline{B}(x_o, r)$ is compact.*

Having squared the table of global aspects, we can now improve on Proposition 1.5, by replacing $d_U(x,y)$ in the right-hand side, with the smaller quantity $d(x,y)$. The price that we must pay is represented by the presence of a larger constant $\tilde{C}$. We stress that, as the above simple example shows, without (1.16) the proof of the next proposition would break down, since the boundedness of the set $\tilde{U}$ would fail in general, see [**G02**].

PROPOSITION 1.10. *Let $U \subset \mathbb{R}^n$ be a bounded set. There exists a bounded set $\tilde{U}$, with $\overline{U} \subset \tilde{U}$, such that for every $x, y \in U$ one has*

$$|x - y| \leq \tilde{C} \, d(x,y).$$

Here,

(1.17) $$\tilde{C} = \left(\max_{z \in \overline{\tilde{U}}} \sum_{j=1}^m |X_j(z)|^2 \right)^{1/2}.$$

This gives for every $x \in U$, and any $r > 0$,

(1.18) $$B(x, r/\tilde{C}) \subset B_e(x, r).$$

1.3. The Nagel-Stein-Wainger polynomial and the size of the CC balls

Let $X = \{X_1,...,X_m\}$ be a system of C^∞ vector fields in $\mathbb{R}^n$, $n \geq 3$, satisfying the finite rank condition (1.4), and denote by $Y_1,...,Y_l$ the collection of the X_j's and of those commutators which are needed to generate $\mathbb{R}^n$. A "degree" is assigned to each Y_i, namely the corresponding order of the commutator. If $I = (i_1,...,i_n), 1 \leq i_j \leq l$, is a n-tuple of integers, following [**NSW84**] one defines $d(I) = \sum_{j=1}^n deg(Y_{i_j})$, and $a_I(x) = \det(Y_{i_1},...,Y_{i_n})$.

DEFINITION 1.11. *The Nagel-Stein-Wainger polynomial is defined by*

$$\Lambda(x,r) = \sum_I |a_I(x)| \, r^{d(I)}, \qquad\qquad r > 0.$$

For a given bounded open set $U \subset \mathbb{R}^n$, we let

(1.19) $\quad Q = \sup\{d(I) \mid |a_I(x)| \neq 0, x \in U\}, \qquad Q(x) = \inf\{d(I) \mid |a_I(x)| \neq 0\},$

and notice that from the work in [**NSW84**] we know
$$3 \leq n \leq Q(x) \leq Q.$$
It is immediate that for every $x \in U$, and every $r > 0$, one has

(1.20) $\qquad t^Q \Lambda(x, r) \leq \Lambda(x, tr) \leq t^{Q(x)} \Lambda(x, r), \qquad 0 \leq t \leq 1.$

The numbers Q and $Q(x)$ are respectively called the *local homogeneous dimension* of U, and the *homogeneous dimension at x*, with respect to the system X. The following fundamental result is due to Nagel, Stein and Wainger [**NSW84**].

THEOREM 1.12. *For every bounded set $U \subset \mathbb{R}^n$ there exist constants $C, R_o > 0$ such that, for any $x \in U$, and $0 < r \leq R_o$, one has*

(1.21) $\qquad\qquad C \, \Lambda(x, r) \leq |B(x, r)| \leq C^{-1} \, \Lambda(x, r).$

As a consequence, with $C_1 = 2^Q$, one has for every $x \in U$, and any $0 < r \leq R_o$

(1.22) $\qquad\qquad |B(x, 2r)| \leq C_1 \, |B(x, r)|.$

Henceforth, the numbers C_1, R_o in (1.22) will be referred to as the *characteristic local parameters* of U with respect to the system X. The doubling condition (1.22) implies

(1.23) $\qquad \left(\dfrac{r}{s}\right)^Q \leq C_1 \dfrac{|B(x_o, r)|}{|B(x_o, s)|}, \qquad x_o \in U, \quad 0 < r < s \leq R_o.$

We will use the following observation [**DGN98**, Cor 2.10].

LEMMA 1.13. *Let $U \subset \mathbb{R}^n$ be a connected, bounded set with $|U| > 0$, and let R_o be as in Theorem 1.12. For any $0 < r \leq R_o$ we have*
$$C_r = \inf_{x \in \overline{U}} |B(x, r)| > 0.$$

In view of Lemma 1.13 we obtain from (1.23) with $C^* = C_1 C_{R_o}^{-1} > 0$

(1.24) $\qquad\qquad \dfrac{r^Q}{|B(x, r)|} \leq C^* R_o^Q \qquad x \in U, \quad 0 < r \leq R_o.$

The following two propositions are easily derived from Theorem 1.12. They will play an important role in the proof of Theorem 8.6.

PROPOSITION 1.14. *The polynomial function $\Lambda(x, r)$ in Definition 1.11 satisfies the following property. Given a bounded set $U \subset \mathbb{R}^n$ one has*
$$Q(x) \frac{\Lambda(x, r)}{r} \leq \frac{\Lambda(x, r_2) - \Lambda(x, r_1)}{r_2 - r_1} \leq Q \frac{\Lambda(x, r)}{r}$$
for any $x \in U$, $0 < r_1 < r_2 < R_o$ and some $r = r(x) \in (r_1, r_2)$. Here, R_o is the characteristic local parameter of U and Q is its local homogeneous dimension (1.19).

Proof. We begin by observing that, from Definition 1.11, for any bounded set $U \subset \mathbb{R}^n$ one has

$$(1.25) \qquad Q(x) \leq \frac{r\Lambda'(x,r)}{\Lambda(x,r)} \leq Q, \qquad \text{for every} \quad x \in U, \quad 0 < r < R_o,$$

where Q and $Q(x)$ are as in (1.19). We fix $x \in U$ and $0 < r_1 < r_2 \leq R_o$, and apply the mean value theorem to the function $\Lambda(x,\cdot)$ to reach the conclusion from (1.25). □

PROPOSITION 1.15. *Let $\alpha \leq n$. For every bounded set $U \subset \mathbb{R}^n$ there exists a constant $C > 0$, depending only on U and X, such that for all $x \in U$, $0 < r_1 \leq r_2 \leq R_o$, one has*

$$\frac{r_2^\alpha}{|B(x,r_2)|} \leq C \frac{r_1^\alpha}{|B(x,r_1)|}.$$

Proof. It is easy to see from Definition 1.11, and from the second equation in (1.19), that for all $x \in U$, $0 < r < R_o$, and $0 \leq t \leq 1$, one has

$$(1.26) \qquad \Lambda(x,tr) \leq t^{Q(x)} \Lambda(x,r).$$

This gives for $0 < r_1 \leq r_2 \leq R_o$, and $\alpha \leq n \leq Q(x)$

$$\frac{r_1^\alpha}{\Lambda(x,r_1)} \geq \left(\frac{r_2}{r_1}\right)^{Q(x)-\alpha} \frac{r_2^\alpha}{\Lambda(x,r_2)} \geq \frac{r_2^\alpha}{\Lambda(x,r_2)}.$$

The conclusion now follows from Theorem 1.12. □

REMARK 1.16. The exponent α in Proposition 1.15 is allowed to be negative. This is important because the limitation on s in Theorem 9.1 depends on the upper and lower bounds of α (see Remark 9.5).

Finally, we recall the following definition from [**NSW84**], p.123. For $x \in \mathbb{R}^n$, and $r > 0$, we set

$$(1.27) \qquad Box(r) = \{x \in \mathbb{R}^n \mid x = \exp\left(\sum_{j=1}^{l} u_j Y_j\right) \text{ with } |u_j| < r^{d_j}\},$$

where we have let $d_j = deg(Y_j)$. Here, exp denotes the exponential mapping associated with the vector fields $Y_1, ..., Y_l$. For its definition and main properties we refer the reader to the appendix of [**NSW84**]. The following result is contained in Theorem 7 in [**NSW84**].

THEOREM 1.17. *Given a bounded set $U \subset \mathbb{R}^n$ there exist $\eta \in (0,1)$, and $R_o > 0$, such that for any $x \in U$, and $0 < r < R_o$, one has*

$$B(x, \eta r) \subset \exp_x(Box(r)) \subset B(x,r).$$

REMARK 1.18. One can be more precise about the shape of the sets $B(x,r)$. They have size r in the directions of the X_j's, whereas they have size r^2 in the directions of the commutators $[X_i, X_j]$, and so on (see [**NSW84**], and also [**Gro96**]).

CHAPTER 2

Carnot groups

It is well-known that the infinitesimal groups naturally associated with a system of smooth vector fields satisfying (1.4) are non-commutative nilpotent Lie groups, whose Lie algebra admits a stratification, see [**St70**], [**F75**], [**RS76**], [**VSC92**], and [**St93**]. These groups, which owe their name to the foundational paper of Charathéodory [**Ca09**] on Carnot thermodynamics, occupy a central position in the study of hypoelliptic partial differential equations, non-commutative harmonic analysis, sub-Riemannian geometry, and CR geometric function theory. Two fundamental results in the subject are the lifting theorem of Rothschild and Stein [**RS76**], and the strong rigidity theorem of Mostow [**Mo73**]. In this chapter we discuss some basic properties of Carnot groups which will play a crucial role in the sequel. A Carnot group G of step r is a simply connected Lie group whose Lie algebra $\mathfrak{g}$ admits a nilpotent stratification of step r. This means that $\mathfrak{g} = V_1 \oplus V_2 \oplus \cdots \oplus V_r$, and that moreover $[V_1, V_j] = V_{j+1}$ for $j = 1, ..., r-1$, whereas $[V_1, V_r] = \{0\}$. We assume that a scalar product $<\cdot, \cdot>$ is given on $\mathfrak{g}$ for which the $V_j's$ are mutually orthogonal. We let $m_j = dim\ V_j$, $j = 1, ..., r$, and denote by

$$N = m_1 + ... + m_r$$

the topological dimension of G. The notation $\{e_{j,1}, ..., e_{j,m_j}\}$, $j = 1, ..., r$, will indicate a fixed orthonormal basis of the $j-th$ layer V_j. Elements of V_j are assigned the formal degree j. As a rule, we will use letters g, g', g_o for points in G, whereas we will reserve the letters ξ, ξ', ξ_o, for elements of the Lie algebra $\mathfrak{g}$. We will denote by

(2.1) $\qquad L_{g_o}(g) = g_o\ g\ , \qquad R_{g_o}(g) = g\ g_o\ ,$

respectively, the left- and right-translations on G by an element $g_o \in G$. Recall that the exponential map $exp : \mathfrak{g} \to G$ is a global analytic diffeomorphism [**V74**]. It allows to define analytic maps $\xi_i : G \to V_i$, $i = 1, ..., r$, by letting $g = \exp(\xi_1(g) + ... + \xi_r(g))$. For $g \in G$, the projection of the *exponential coordinates* of g onto the layer V_j, $j = 1, ..., r$, are defined as follows

(2.2) $\qquad x_{j,s}(g) = <\xi_j(g), e_{j,s}>, \qquad s = 1, ..., m_j.$

It will be convenient to have a separate notation for the first two layers V_1 and V_2. For simplicity, we set $m = m_1$, $k = m_2$, and indicate with

(2.3) $\quad \{e_1, ...e_m\} = \{e_{1,1}, ..., e_{1,m_1}\}\ , \qquad \{\epsilon_1, ..., \epsilon_k\} = \{e_{2,1}, ..., e_{2,m_2}\}\ ,$

respectively the orthonormal basis of V_1 and V_2. We denote by $X = \{X_1, ..., X_m\}$ and $Y = \{Y_1, ..., Y_k\}$ the corresponding systems of left-invariant vector fields on G

defined by

$$X_j(g) = (L_g)_*(e_j), j = 1,...,m, \qquad Y_l(g) = (L_g)_*(\epsilon_l), l = 1,...,k,$$

where $(L_g)_*$ denotes the differential of L_g. The system X defines a basis for the so-called *horizontal sub-bundle* $H\mathbf{G}$ of the tangent bundle $T\mathbf{G}$. For a given function $f : \mathbf{G} \to \mathbb{R}$, the action of X_j on f is specified by the equation

$$(2.4) \qquad X_j f(g) = \lim_{t \to 0} \frac{f(g \, exp \, (tX_j)) - f(g)}{t} = \frac{d}{dt} f(g \, exp \, (tX_j))\big|_{t=0}.$$

A similar formula holds for any left-invariant vector field. We indicate with
$$(2.5) \qquad x_j(g) = \, <\xi_1(g), e_j>, \quad j = 1,...,m, \qquad y_s(g) = \, <\xi_2(g), \epsilon_s>, \quad s = 1,...,k.$$
the projections of the exponential coordinates of g onto V_1 and V_2. Letting $x(g) = (x_1(g),...,x_m(g))$, $y(g) = (y_1(g),...,y_k(g))$, we will often identify $g \in \mathbf{G}$ with its exponential coordinates

$$(2.6) \qquad g = (x(g), y(g), ...),$$

where the dots indicate the $(N - (m + k))$-dimensional vector

$$(x_{3,1}(g),...,x_{3,m_3}(g),...,x_{r,1}(g),...,x_{r,m_r}(g)).$$

When $\mathbf{G}$ is a group of step 2, then (2.6) simply becomes $g = (x(g), y(g))$. Such identification of $\mathbf{G}$ with its Lie algebra is justified by the Baker-Campbell-Hausdorff formula, see, e.g., [**V74**]

$$(2.7) \qquad exp \, \xi \, exp \, \xi' = exp \, (\xi + \xi' + \frac{1}{2} [\xi, \xi'] + ...) \qquad\qquad \xi, \xi' \in \mathfrak{g},$$

where the dots indicate a finite linear combination of terms containing commutators of order two and higher. For $\xi \in \mathfrak{g}$ consider the map $\theta_\xi : \mathfrak{g} \to \mathfrak{g}$ given by

$$(2.8) \qquad \theta_\xi(\xi') = \xi + \xi' + \frac{1}{2} [\xi, \xi'] + ...$$

where the right-hand side is given by the Baker-Campbell-Hausdorff sum in (2.7). If we endow the Lie algebra $\mathfrak{g}$ with the polynomial group law

$$(2.9) \qquad \xi \circ \xi' = \theta_\xi(\xi'),$$

then we can identify the group $\mathbf{G}$ with $\mathfrak{g}$, via the exponential coordinates. In a Carnot group one has $X_j^* = -X_j$ [**F75**]. The *sub-Laplacian* associated with a basis X is the second-order partial differential operator on $\mathbf{G}$ given by

$$(2.10) \qquad \mathcal{L} = \sum_{j=1}^m X_j^2 = -\sum_{j=1}^m X_j^* X_j.$$

By the assumption on the Lie algebra one immediately sees that the system X satisfies the finite rank condition (1.4), therefore thanks to Hörmander's theorem [**H67**] the operator $\mathcal{L}$ is hypoelliptic. Every Carnot group is naturally equipped with a family of non-isotropic dilations. One first defines dilations $\Delta_\lambda : \mathfrak{g} \to \mathfrak{g}$ on the Lie algebra as follows. If $\xi = \xi_1 + ... + \xi_r \in \mathfrak{g}$, with $\xi_j \in V_j$, $j = 1,...,r$, one lets

$$(2.11) \qquad \Delta_\lambda \, \xi = \Delta_\lambda(\xi_1 + \cdots + \xi_r) = \lambda \, \xi_1 + \cdots + \lambda^r \, \xi_r.$$

One then uses the exponential mapping to lift (2.11) to the group, i.e.,

$$(2.12) \qquad \delta_\lambda(g) = exp \circ \Delta_\lambda \circ exp^{-1}(g), \qquad g \in \mathbf{G}.$$

Denoting by dg the bi-invariant Haar measure on $\boldsymbol{G}$ obtained by lifting via the exponential map exp the Lebesgue measure on $\mathfrak{g}$, one easily checks that

$$(d \circ \delta_\lambda)(g) = \lambda^Q \, dg, \qquad \text{where} \quad Q = \sum_{j=1}^{r} j \, dim(V_j).$$

The number Q, called the *homogeneous dimension* of G, plays an important role in the analysis of Carnot groups. In the non-abelian case $r > 1$, one clearly has $Q > N$.

We denote by $d(g, g')$ the *CC distance* on $\boldsymbol{G}$ associated with the system X. It is well-known that $d(g, g')$ is equivalent to the *gauge pseudo-metric* $\rho(g, g')$ on $\boldsymbol{G}$, i.e., there exists a constant $C = C(\boldsymbol{G}) > 0$ such that

(2.13) $\qquad C \, \rho(g, g') \leq d(g, g') \leq C^{-1} \, \rho(g, g'), \qquad g, g' \in \boldsymbol{G},$

see [**NSW84**], [**VSC92**]. The pseudo-distance $\rho(g, g')$ is defined as follows. Let $|\cdot|$ denote the Euclidean distance to the origin on $\mathfrak{g}$. For $\xi = \xi_1 + \cdots + \xi_r \in \mathfrak{g}$, $\xi_i \in V_i$, one lets

(2.14) $\qquad |\xi|_{\mathfrak{g}} = \left(\sum_{i=1}^{r} |\xi_i|^{2r!/i} \right)^{2r!}, \qquad |g|_{\boldsymbol{G}} = |\exp^{-1} g|_{\mathfrak{g}}, \qquad g \in \boldsymbol{G}.$

The pseudo-distance on $\boldsymbol{G}$ associated to $|\cdot|_{\boldsymbol{G}}$ is given by

(2.15) $\qquad \rho(g, g') = |g^{-1} g'|_{\boldsymbol{G}}.$

Denoting with
(2.16)
$\qquad B(g, R) = \{g' \in \boldsymbol{G} \mid d(g', g) < R\}, \qquad B_\rho(g, R) = \{g' \in \boldsymbol{G} \mid \rho(g', g) < R\},$

respectively the CC ball and the gauge pseudo-ball centered at g with radius R, one easily recognizes that there exist $\omega = \omega(\boldsymbol{G}) > 0$, and $\alpha = \alpha(\boldsymbol{G}) > 0$ such that

(2.17) $\qquad |B(g, R)| = \omega \, R^Q, \qquad |B_\rho(g, R)| = \alpha \, R^Q, \qquad g \in \boldsymbol{G}, R > 0.$

The first equation in (2.17) shows, in particular, that for a Carnot group the Nagel-Stein-Wainger polynomial in Definition 1.11 is simply the monomial ωR^Q.

2.1. Carnot groups of step 2

A class of Carnot groups of special geometric interest is that of groups of step 2. Let $\boldsymbol{G}$ be such a group, with Lie algebra $\mathfrak{g} = V_1 \oplus V_2$, and denote by b_{ij}^l the *group constants* defined by the equation

(2.18) $\qquad [e_i, e_j] = \sum_{l=1}^{k} b_{ij}^l \, \epsilon_l,$

where $\{e_1, ..., e_m\}$ and $\{\epsilon_1, ..., \epsilon_k\}$ are as in (2.3). Using (2.7), one immediately sees that in the exponential coordinates

$$Y_l = \frac{\partial}{\partial y_l}, \qquad l = 1, ..., k \, .$$

The following useful formula for the derivative along the vector fields X_j in exponential coordinates holds.

LEMMA 2.1. *Let $f : \mathbf{G} \to \mathbb{R}$, then*

$$X_j f(g) = \frac{\partial f}{\partial x_j}(g) + \frac{1}{2} \sum_{l=1}^{k} \left(\sum_{i=1}^{m} b_{ij}^l x_i(g) \right) \frac{\partial f}{\partial y_l}(g).$$

Proof. To prove the lemma we recall the definition (2.4) of $X_j f(g)$. Let $g = \exp \xi(g)$, with $\xi(g) = \xi_1(g) + \xi_2(g)$. Using (2.7) one obtains

$$g \exp (te_j) = \exp \left(\xi_1(g) + te_j + \xi_2(g) + \frac{t}{2} [\xi_1(g), e_j] \right).$$

From (2.18) we find

$$[\xi_1(g), e_j] = \sum_{l=1}^{k} \left(\sum_{i=1}^{m} b_{ij}^l x_i(g) \right) \epsilon_l,$$

and therefore

$$f(g \exp (te_j))$$
$$= f\Big(x_1(g), ..., x_j(g) + t, ..., x_m(g), y_1(g)$$
$$+ \frac{t}{2} \sum_{i=1}^{m} b_{ij}^1 x_i(g), ..., y_k(g) + \frac{t}{2} \sum_{i=1}^{m} b_{ij}^k x_i(g)\Big).$$

Differentiating the latter equation with respect to t, and setting $t = 0$, we obtain the conclusion. □

For $\xi \in \mathfrak{g}$ consider the map $\theta_\xi : \mathfrak{g} \to \mathfrak{g}$ defined by (2.8). In view of (2.7), θ_ξ is a Lie algebra homomorphism.

PROPOSITION 2.2. *Let $\xi' = \xi'_1 + \xi'_2 \in \mathfrak{g}$, with $\xi'_1 \in V_1, \xi'_2 \in V_2$, then $\theta_{\xi'}$ is an affine transformation whose Jacobian is given by*

$$(2.19) \qquad d\theta_{\xi'} = \begin{pmatrix} Id_{m \times m} & 0_{m \times k} \\ J_{k \times m} & Id_{k \times k} \end{pmatrix}.$$

Here, J is a $k \times m$ matrix with entries

$$J(l, j) = \frac{1}{2} \sum_{i=1}^{m} b_{i,j}^l x'_i, \qquad 1 \leq l \leq k, \quad 1 \leq j \leq m.$$

Proof. Let $\xi = \xi_1 + \xi_2$, then $[\xi', \xi] = [\xi'_1, \xi_1]$, and we obtain from (2.18)

$$(2.20) \qquad \theta_{\xi'}(\xi) = \xi' + \xi + \frac{1}{2} \sum_{i,j=1}^{m} x'_i x_j [e_i, e_j]$$

$$= \xi' + \xi + \frac{1}{2} \sum_{l=1}^{k} \left(\sum_{i,j=1}^{m} b_{i,j}^l x'_i x_j \right) \epsilon_l.$$

The conclusion follows immediately from the last expression. □

2.2. The Kaplan mapping

In a group G of step 2, with Lie algebra $\mathfrak{g} = V_1 \oplus V_2$, consider the linear mapping $J : V_2 \to End(V_1)$ defined by

$$(2.21) \qquad < J(\eta)\xi', \xi'' > \; = \; < [\xi', \xi''], \eta >, \qquad \eta \in V_2, \quad \xi', \xi'' \in V_1.$$

The algebraic properties of the mapping J have important repercussions on the geometric and analytic properties of Carnot groups of step 2. An immediate consequence of the definition of J is that

$$(2.22) \qquad < J(\eta)\xi, \xi > \; = \; 0, \qquad \text{for every} \quad \eta \in V_2, \; \xi \in V_1.$$

LEMMA 2.3. *In a Carnot group G of step 2, consider the function $\psi(g) = |x(g)|^2$. For any $s = 1, ..., k$, one has*

$$(2.23) \qquad < X\psi, Xy_s > \; \equiv \; 0.$$

Let $l = 1, ..., k$ be fixed, and denote by $y'(g)$ the $(k-1)$-dimensional vector obtained from $y(g)$, by removing the component $y_l(g)$. One has

$$(2.24) \qquad < X\psi, X(|y'|^2) > \; = \; 0.$$

Proof. Let $g = \exp \xi$, with $\xi = \xi_1 + \xi_2$. For $t \in \mathbb{R}$, one has from (2.7)

$$(2.25) \qquad y_s(g \, \exp t e_j) \; = \; y_s(g) + \frac{t}{2} < [\xi_1(g), e_j], \epsilon_s >$$
$$= \; y_s(g) + \frac{t}{2} < J(\epsilon_s)\xi_1(g), e_j > .$$

We have from (2.25)

$$(2.26) \qquad (X_j \, y_s)(g) \; = \; \frac{1}{2} < J(\epsilon_s)\xi_1(g), e_j > .$$

On the other hand, we easily obtain from Lemma 2.1

$$(2.27) \qquad X_j \psi(g) \; = \; 2 x_j(g) \; = \; 2 < \xi_1(g), e_j > .$$

Equations (2.26), (2.27) give

$$< X\psi, Xy_s > (g) \; = \; \sum_{j=1}^{m} X_j \psi(g) \, (X_j \, y_s)(g) \; = \; \sum_{j=1}^{m} < \xi_1(g), e_j > < J(\epsilon_s)\xi_1(g), e_j >$$
$$= \; < J(\epsilon_s)\xi_1(g), \xi_1(g) > \; = \; 0,$$

where in the last equality we have used (2.22). This proves (2.23). From the latter, (2.24) immediately follows, since

$$< X\psi, X(|y'|^2) > \; = \; 2 \sum_{\substack{s=1 \\ s \neq l}}^{k} y_s < X\psi, X(y_s) > \; = \; 0.$$

□

2.3. Groups of Heisenberg type

We next recall an important class of Carnot groups of step 2 which is modelled on the Heisenberg group $\mathbb{H}^n$, but whose geometry is more intricated since the center can have arbitrary dimension. The introduction of such groups is due to A. Kaplan [**K80**], [**K81**], [**K83**].

DEFINITION 2.4. *A Carnot group G of step 2 is called of Heisenberg type if for every $\eta \in V_2$, such that $|\eta| = 1$, the map $J(\eta) : V_1 \to V_1$ is orthogonal.*

We stress that there exists a plentiful supply of groups of Heisenberg type. For instance, the nilpotent component N in the Iwasawa decomposition $G = KAN$, where G is a simple group of rank one, is a group of Heisenberg type [**CDKR91**]. Such groups N are called *Iwasawa groups*. When the center V_2 of the group is one-dimensional, then (up to isomorphisms) a group of Heisenberg type is nothing but the Heisenberg group $\mathbb{H}^n$. Because of their symmetries, groups of Heisenberg type play a distinguished role in analysis and geometry. For either aspect the reader can consult the following (non-exaustive) list of references: [**F73**], [**FS74**], [**Gav77**], [**Ge77**], [**Ge79**], [**Ge(I)80**], [**Ge(II)80**], [**Ge(III)80**], [**Gre80**], [**Gre81**], [**J81**], [**Cy81**], [**GeS82**], [**P82**], [**Ko83**], [**KaR83**], [**CK84**], [**JL84**], [**GGV84**], [**Ge84**], [**GeS84**], [**Ko85**], [**Ko85**], [**KoR85**], [**KoS85**], [**Da85**], [**BGGV86**], [**GGV86**], [**Da87**], [**Da87**], [**KoR87**], [**JL87**], [**JL88**], [**JL89**], [**BG88**], [**F89**], [**HM89**], [**Ge90**], [**GL90**], [**GS90**], [**Mu90**], [**MR90**], [**Th90**], [**ABCP91**], [**CDKR91**], [**Ro91**], [**Th(I)91**], [**Th(II)91**], [**Th(III)91**], [**Th(IV)91**], [**HoR92**], [**MR92**], [**ABC93**], [**Me93**], [**St93**], [**Th93**], [**ABC94**], [**ABCP94**], [**E(I)94**], [**E(II)94**], [**MS94**], [**Th94**], [**KoR95**], [**MRS95**], [**SST95**], [**Th95**], [**CDG96**], [**Gro96**], [**E96**], [**MRS96**], [**CMZ96**], [**RT96**], [**MPR97**], [**NT97**], [**RRT97**], [**CG98**], [**LU98**], [**BGJS98**], [**CDKR**], [**Ge98**], [**Th98**], [**AR99**], [**BPr99**], [**MS99**], [**BGG00**], [**Bi00**], [**GV00**], [**Pa00**], [**Th00**], [**FSS01**], [**GV01**], [**NT01**], [**DGN02**], [**CGN02**], [**CG03**], [**DGN03(I)**], [**DGN03(II)**], [**Pa(I)04**], [**Pa(II)04**], [**DGN04(I)**], [**DGN04(II)**], [**CGP04**], [**GM04**], [**GT04**], [**GP04**], [**CHMY04**].

Definition 2.4 implies

$$|J(\eta)\xi| = |\eta|\,|\xi|, \qquad \eta \in V_2, \quad \xi \in V_1, \tag{2.28}$$

$$<J(\eta')\xi, J(\eta'')\xi> = <\eta',\eta''>|\xi|^2, \qquad \eta',\eta'' \in V_2,\ \xi \in V_1. \tag{2.29}$$

In the next lemma we establish some key properties of groups of Heisenberg type.

LEMMA 2.5. *Let G be a group of Heisenberg type. For any fixed $l = 1, ..., k$, one has*

$$<X(y_l)(g), X(|y'|^2)(g)> = 0. \tag{2.30}$$

$$|X(y_l)(g)|^2 = \frac{1}{4}|x(g)|^2. \tag{2.31}$$

$$|X(|y'|^2)(g)|^2 = |x(g)|^2\,|y'(g)|^2. \tag{2.32}$$

Proof. One has from (2.26)

(2.33)
$$< X(y_l)(g), X(|y'|^2)(g) > = 2 \sum_{\substack{s=1 \\ s \neq l}}^{k} y_s \sum_{j=1}^{m} X_j(y_l) X_j(y_s)$$

$$= \frac{1}{2} \sum_{\substack{s=1 \\ s \neq l}}^{k} y_s \sum_{j=1}^{m} < J(\epsilon_l)\xi_1, e_j >< J(\epsilon_s)\xi_1, e_j >$$

$$= \frac{1}{2} \sum_{\substack{s=1 \\ s \neq l}}^{k} y_s < J(\epsilon_l)\xi_1, J(\epsilon_s)\xi_1 >$$

It is at this point that we use the Heisenberg type structure of $\boldsymbol{G}$, obtaining from (2.29)
$$< J(\epsilon_l)\xi_1, J(\epsilon_s)\xi_1 > = < \epsilon_l, \epsilon_s > |\xi_1|^2 = \delta_{ls} |x|^2.$$

Substituting the latter equation in (2.33), we conclude
$$< X(y_l), X(|y'|^2) > = \frac{1}{2} \left(\sum_{\substack{s=1 \\ s \neq l}}^{k} y_s \delta_{sl} \right) |x|^2 = 0.$$

This proves (2.30). To establish (2.31), we use (2.26) and (2.28) which give
$$|X(y_l)|^2 = \frac{1}{4} \sum_{j=1}^{m} < J(\epsilon_l)(\xi_1), e_j >^2 = \frac{1}{4} |J(\epsilon_l)(\xi_1)|^2 = \frac{1}{4} |\epsilon_l|^2 |\xi_1|^2 = \frac{1}{4} |x|^2.$$

Finally, again from (2.26), we obtain
$$X_j(|y'|^2) = 2 \sum_{\substack{s=1 \\ s \neq l}}^{k} y_s X_j(y_s) = \sum_{\substack{s=1 \\ s \neq l}}^{k} y_s < J(\epsilon_s)(\xi_1), e_j > = < [\xi_1, e_j], \eta' >,$$

where we have let $\eta' = \sum_{\substack{s=1 \\ s \neq l}}^{k} y_s \epsilon_s$. This formula gives
$$|X(|y'|^2)|^2 = \sum_{j=1}^{m} (X_j(|y'|^2))^2 = \sum_{j=1}^{m} < [\xi_1, e_j], \eta' >^2$$

$$= \sum_{j=1}^{m} < J(\eta')(\xi_1), e_j >^2 = |J(\eta')\xi_1|^2.$$

If at this point we use (2.28), we conclude
$$|X(|y'|^2)|^2 = |J(\eta')\xi_1|^2 = |\eta'|^2 |\xi_1|^2 = |x|^2 |\eta'|^2.$$

This establishes (2.32), and completes the proof. $\square$

We close this chapter by recalling an important formula due to Kaplan [**K80**], which generalized a basic discovery of Folland for the Heisenberg group [**F73**]. In a group of Heisenberg type $\boldsymbol{G}$ we consider the renormalized gauge

(2.34)
$$N(g) = \left(|x(g)|^4 + 16|y(g)|^2 \right)^{1/4}.$$

We notice that (2.34) differs from the expression given by the general formula (2.14) in the case $r = 2$ only for the (immaterial) normalization factor 16. Let $\mathcal{L}$ be a sub-Laplacian associated with an orthonormal basis X of the first layer of the Lie algebra of $\boldsymbol{G}$, and denote by $\Gamma(g, g')$ the corresponding positive fundamental solution. There exists $C(\boldsymbol{G}) > 0$ such that

$$\Gamma(g, g') = \frac{C(\boldsymbol{G})}{\rho(g, g')^{Q-2}} \qquad g, g' \in \boldsymbol{G}, g \neq g', \tag{2.35}$$

where $\rho(g, g') = N(g^{-1}g')$.

CHAPTER 3

The characteristic set

The notion of characteristic set is central to the subjects of sub-elliptic equations and of CC geometry. In this chapter we analyze it in detail and discuss various geometric situations of interest. We begin by recalling that an open set $\Omega \subset \mathbb{R}^n$ is said to be of class C^1 if for every $x_o \in \partial\Omega$ there exist a neighborhood U_{x_o} of x_o, and a function $\phi_{x_o} \in C^1(U_{x_o})$, with $|\nabla \phi_{x_o}| \geq \alpha > 0$ in U_{x_o}, such that
(3.1)
$$\Omega \cap U_{x_o} = \{x \in U_{x_o} \mid \phi_{x_o}(x) < 0\}, \qquad \partial\Omega \cap U_{x_o} = \{x \in U_{x_o} \mid \phi_{x_o}(x) = 0\}.$$

DEFINITION 3.1. Let $\Omega \subset \mathbb{R}^n$ be an open set of class C^1. A point $x_o \in \partial\Omega$ is called *characteristic* with respect to the system X, if given U_{x_o}, ϕ_{x_o}, as in (3.1), one has $X_1 \phi_{x_o}(x_o) = 0, ..., X_m \phi_{x_o}(x_o) = 0$, or equivalently
(3.2)
$$|X\phi_{x_o}(x_o)| = 0.$$
The *characteristic set* $\Sigma = \Sigma_{\Omega,X}$ is the collection of all characteristic points of Ω with respect to X.

We notice that the condition (3.2) in Definition 3.1 is equivalent to the more familiar one
$$\Sigma = \Sigma_{\Omega,X} = \{x_o \in \partial\Omega \mid X_j(x_o) \in T_{x_o}(\partial\Omega), \quad j = 1, ..., m\}.$$
The *angle function* of Ω is defined by
(3.3)
$$w(x) \stackrel{def}{=} |X\phi(x)|, \qquad x \in \partial\Omega.$$
The reason for the name is in the fact that $w(x)$ measures the angle formed by the outer unit normal $\nu(x)$ to $\partial\Omega$ at the point x with span $\{X_1(x), ..., X_m(x)\}$. In fact, since the latter is given by $\nu = \nabla\phi/|\nabla\phi|$, it is clear that if $x \in \partial\Omega$, then one has
$$\left\{ \sum_{j=1}^{m} <X_j(x), \nu(x)>^2 \right\}^{1/2} = \frac{\left\{ \sum_{j=1}^{m} X_j\phi(x)^2 \right\}^{1/2}}{|\nabla\phi(x)|} = \frac{w(x)}{|\nabla\phi(x)|}.$$

3.1. A result of Derridj on the size of the characteristic set

Henceforth, we denote by H_s the s-dimensional Hausdorff measure in $\mathbb{R}^n$ constructed with the standard Euclidean distance, see, e.g., [**Fe69**]. If $D \subset \mathbb{R}^n$ is a C^1, or a Lipschitz domain, then $H_{n-1} \lfloor \partial D$ is just the ordinary surface measure on ∂D. Suppose that $\phi : \mathbb{R}^n \to \mathbb{R}$ be a C^1 defining function for Ω, i.e.,

$\Omega = \{x \in \mathbb{R}^n \mid \phi(x) < 0\}$. Denoting with ∇ the standard gradient in $\mathbb{R}^n$, we always assume that $|\nabla \phi(x)| > 0$, for every $x \in \partial \Omega$. In particular, when Ω is a bounded set we infer the existence of constants $\beta_\Omega \geq \alpha_\Omega$, such that

$$(3.4) \qquad 0 < \alpha_\Omega \leq |\nabla \phi(x)|^{-1} \leq \beta_\Omega, \qquad \text{for every} \quad x \in \partial \Omega.$$

Typically, bounded domains have non-empty characteristic sets. For instance, due to topological reasons every bounded C^1 domain in the Heisenberg group $\mathbb{H}^n$, whose boundary is homeomorphic to the $2n$-dimensional sphere $\mathbb{S}^{2n}$, has non-empty characteristic set. The following basic result, due to Derridj [**De71**], [**De72**], shows that, at least from the measure theoretic point of view, the set Σ is not too big.

THEOREM 3.2. *Let $\Omega \subset \mathbb{R}^n$ be a C^∞ domain. One has*

$$H_{n-1}(\Sigma_{\Omega,X}) = 0.$$

Although we will not use it in the present paper, in connection with the size of the characteristic set we mention the interesting recent results of Balogh and Magnani. The former author [**B00**] has proved that for a C^1 domain in the Heisenberg group $\mathbb{H}^n$, the characteristic set has zero $(Q-1)$-dimensional Hausdorff measure with respect to the CC distance of the group. Magnani has extended Balogh's theorem to Carnot groups of step 2 in [**M03(I)**], and subsequently to groups of arbitrary step [**M03(II)**].

3.2. Some geometric examples

In this chapter we discuss various examples of characteristic sets which have a special geometric interest in the applications. Typically, in the theory of sub-elliptic equations, or in CC geometry, characteristic points are present. In the sense that a domain with an empty characteristic set must possess some special properties, either geometric or topological. We will illustrate these aspects with some examples.

3.3. Non-characteristic manifolds

Our goal here is to provide examples of domains whose boundary has empty characteristic set. We begin with the case of unbounded domains. As we will see, the construction of bounded domains is much more delicate, and it involves topology.

EXAMPLE 3.3 (**Non-characteristic hyper-planes**). Let $\boldsymbol{G}$ be a Carnot group of step r, with Lie algebra $\mathfrak{g} = V_1 \oplus ... \oplus V_r$, and let $m = dim(V_1)$. For a fixed vector $a \in \mathbb{R}^m \setminus \{0\}$, and for $\lambda \in \mathbb{R}$, consider the *half-space*

$$H_a^+ = \{g \in \boldsymbol{G} \mid < x(g), a >\, > \lambda\}.$$

One has $\Sigma = \Sigma_{H_a^+, X} = \varnothing$. Furthermore, the relative angle function $w = |X\phi|$, where $\phi(g) = \lambda - < x(g), a >$, satisfies the equation $w(g) \equiv |a|$, for every $g \in \boldsymbol{G}$.

Proof. Using (2.7) we easily see that for every fixed $i = 1, ..., m$ one has

$$(3.5) \qquad x_j(g \exp te_i) = x_j(g) + t\, \delta_{ij}, \qquad j = 1, ..., m.$$

The latter equation shows that

$$|X\phi(g)|^2 = \sum_{i=1}^{m}(X_i\phi(g))^2 = |a|^2.$$

This proves the above claims. $\square$

The above example, and variants of it, are of interest in the study of the CR Yamabe problem, see [**LU98**], [**GV00**], and [**GV01**], and also in the theory of minimal surfaces in Carnot groups [**FSS01**], [**DGN04(II)**]. When the domain Ω is bounded, then topology enters the picture, and the construction of non-characteristic boundaries is a much less obvious task. We provide a significant model in the following example.

EXAMPLE 3.4 (**A toroid in a group of Heisenberg type G**).

To describe such set consider a Carnot group G of step 2, with Lie algebra $\mathfrak{g} = V_1 \oplus V_2$. Let M be a k-dimensional compact submanifold of the $(k+1)$-dimensional vector space

$$\{\xi \in V_1 \mid <\xi, e_2> = ... = <\xi, e_m> = 0\} \times V_2 \subset \mathfrak{g}.$$

We assume in addition that $\partial M = \varnothing$, and that

$$d_e(M, \{0\} \times V_2) > 0,$$

where $d_e(M, \{0\} \times V_2)$ denotes the Euclidean distance in $V_1 \times V_2$. If $\mathbb{S}^{m-1}$ denotes the $(m-1)$-dimensional sphere in V_1 centered at the origin, we define the *toroid* generated by M as

(3.6) $$\mathbb{T}(M) = \{g \in G \mid \exp^{-1}(g) \in \mathbb{S}^{m-1} \times M\}.$$

We want to establish the following.

PROPOSITION 3.5. *Let $\mathbb{T}(M)$ be a toroid in a group of Heisenberg type, then*

$$\Sigma = \Sigma_{\mathbb{T}(M),X} = \varnothing.$$

Before presenting the proof of Proposition 3.5 we need to develop some preliminary material. The following definition is taken from [**GV01**].

DEFINITION 3.6. Let G be a Carnot group of step 2 with Lie algebra $\mathfrak{g} = V_1 \oplus V_2$. A bounded domain $\Omega \subset G$ is said to have *partial symmetry* if, for some $g_o \in G$, the domain $L_{g_o}(\Omega)$ is invariant under the action of the orthogonal group $\mathbb{O}(m)$ onto the first layer V_1.

Here, $L_{g_o} : G \to G$ is the operator of left-translation introduced in (2.1). We remark that the toroid in (3.6) has partial symmetry (with respect to the group identity e). Our next goal is to understand the location of the characteristic set of a domain with partial symmetry. It is clear that, since the condition that a point be characteristic is invariant under the left-translations $\{L_g\}_{g \in G}$, it will suffice to look at the situation in which $g_o = e$. By left-translating along the center, we can without restriction assume that $e \in \partial\Omega$. In what follows, we will consider an even more general situation, and suppose in fact that, in a neighborhood V of e, the domain Ω can be described as follows: There exists a fixed index $l = 1, ..., k$, such that

(3.7) $\quad\quad\quad \partial\Omega \cap V = \{g \in \boldsymbol{G} \mid y_l(g) = f(|x(g)|^4, |y'(g)|^2)\} \cap V,$

where $f : \mathbb{R}^2 \to \mathbb{R}$ is a given C^1 function, with $f(0,0) = 0$. If we let

(3.8) $\quad\quad\quad\quad\quad\quad \phi(g) = y_l(g) - f(|x(g)|^4, |y'(g)|^2),$

then ϕ is a defining function for Ω near e. We want to compute the relative angle function. Letting $t_1 = t_1(g) = |x(g)|^4$, and $t_2 = t_2(g) = |y'(g)|^2$, we find

$$X_j\phi = X_j y_l - \left[2 f_{t_1}(|x|^4, |y'|^2) \psi X_j\psi + f_{t_2}(|x|^4, |y'|^2) X_j(|y'|^2)\right],$$

where for simplicity we have omitted the argument g for all functions involved. The latter equation gives

(3.9) $\quad |X\phi|^2 = |X(y_l)|^2 + 4 f_{t_1}^2 \psi^2 |X\psi|^2 + f_{t_2}^2 |X(|y'|^2)|^2$
$\quad\quad\quad\quad + 4 f_{t_1} f_{t_2} \psi <X\psi, X(|y'|^2)> - 4 f_{t_1} \psi <X(y_l), X\psi>$
$\quad\quad\quad\quad - 2 f_{t_2} <X(y_l), X(|y'|^2)>.$

Lemma 2.3 allows to obtain from (3.9)

(3.10) $\quad |X\phi|^2 = |X(y_l)|^2 + 16 f_{t_1}^2 \psi^3 + f_{t_2}^2 |X(|y'|^2)|^2$
$\quad\quad\quad\quad - 2 f_{t_2} <X(y_l), X(|y'|^2)>.$

To understand further the nature of the right-hand side in (3.10), we need to impose more conditions on the geometry of the group $\boldsymbol{G}$. An important class of Carnot groups of step 2 for which, in most occasions, a better understanding of the angle function can be obtained is that of groups of Heisenberg type. With Lemma 2.5 in hands we can now complete the identification of the characteristic set of a domain described by (3.7).

PROPOSITION 3.7. *In a group of Heisenberg type $\boldsymbol{G}$, consider a bounded C^1 domain Ω such that, for some $l = 1, ..., k$, one can describe $\partial\Omega$ as in (3.7). One has*

$$\Sigma = \{g \in \partial\Omega \mid x(g) = 0\}.$$

Proof. Using Lemma 2.5 in (3.10) we obtain

(3.11) $\quad |X\phi(g)|^2 = \frac{1}{4} |x(g)|^2 + 16 f_{t_1}^2 |x(g)|^6 + f_{t_2}^2 |x(g)|^2 |y'(g)|^2$
$\quad\quad\quad\quad = |x(g)|^2 \left\{\frac{1}{4} + 16 f_{t_1}^2 |x(g)|^4 + f_{t_2}^2 |y'(g)|^2\right\}.$

From (3.11) we deduce that $|X\phi(g)| = 0$ if and only if $x(g) = 0$. This proves the proposition. $\quad\square$

After this preliminary work, we can finally prove Proposition 3.5.

Proof of Proposition 3.5. From its definition it is clear that the compact manifold $\mathbb{T}(M)$ is invariant under the action of the orthogonal group $\mathbb{O}(m)$ on the first layer V_1. Therefore, as a special case of Proposition 3.7, we deduce that the only characteristic points of $\mathbb{T}(M)$ can occur on the set $\mathcal{A} = \mathbb{T}(M) \cap \{g \in \boldsymbol{G} \mid x(g) = 0\}$. However, by the assumption $d_e(M, \{0\} \times V_2) > 0$, we conclude that $\mathcal{A} = \emptyset$. This completes the proof. $\quad\square$

3.4. Manifolds with controlled characteristic set

The previous two examples provide instances in which the characteristic set is empty. We next discuss several geometrically interesting examples in which the characteristic set is non-empty, and perhaps quite large, but nonetheless can be controlled in an appropriate sense. We introduce a specific definition.

DEFINITION 3.8. Let $X = \{X_1, ..., X_m\}$ be a system of C^∞ vector fields in $\mathbb{R}^n$ satisfying the finite rank condition (1.4). Given a C^1 open set $\Omega \subset \mathbb{R}^n$, we say that its characteristic set $\Sigma = \Sigma_{\Omega,X}$ is *controlled* if there exists a constant $C > 0$, such that for every $x_o \in \Sigma$, with local defining function $\phi_{x_o} : U_{x_o} \to \mathbb{R}$, one has for every $x \in \partial\Omega \cap U_{x_o}$

$$(3.12) \qquad w(x) = |X\phi(x)| \geq C\, d(x, \Sigma).$$

In (3.12) the distance is measured with respect to the CC metric $d(x,y)$ associated with X, see chapter 1.1. It is important to observe that, if Ω is a bounded domain with empty characteristic set (see Example 2 in chapter 3.3), then there exists a constant $C = C(\Omega, X) > 0$ such that $w(x) \geq C$ for every $x \in \partial\Omega$. Therefore, in this situation the condition (3.12) is trivially fulfilled. Definition 3.8 is connected with the notion of *strongly isolated characteristic point*, introduced by D. Jerison [**J81**], but it is a weaker condition, see Example 5 below. Its relevance is due to the fact that when the characteristic set is controlled, then it is easier to study its properties. In the sequel we will construct several basic examples of sets whose characteristic set is non-empty, but it is controlled.

EXAMPLE 3.9 (**Gauge balls in a group of Heisenberg type**).

PROPOSITION 3.10. *In a group of Heisenberg type $\boldsymbol{G}$ the characteristic set of a gauge ball*

$$B_\rho(g_o, r) = \{g' \in \boldsymbol{G} \mid \rho(g_o, g') < r\}$$

is controlled.

Proof. By left-translation and dilation we can assume that $g_o = e$, the group indentity, and that $r = 1$, so that ∂B_ρ, where $B_\rho = B_\rho(e,1)$, is described by the equation

$$(3.13) \qquad |x(g)|^4 + 16\, |y(g)|^2 = 1,$$

see (2.34). A defining function ϕ for B_ρ is given by

$$\phi(g) = \rho(g)^4 - 1 = |x(g)|^4 + 16\, |y(g)|^2 - 1 = \psi(g)^2 + 16\,|y(g)|^2 - 1.$$

Since

$$X\phi = 2\,\psi\, X\psi + 16\, X(|y|^2),$$

we thus obtain

$$|X\phi|^2 = 4\,\psi^2\, |X\psi|^2 + 16^2\, |X(|y|^2)|^2 + 64\,\psi\, <X\psi, X(|y|^2)>.$$

From Lemma 2.1, Lemma 2.3, and from the proof of (2.32) we conclude

$$(3.14) \qquad |X\phi|^2 = 16\,|x|^2\,(|x|^4 + 16\,|y|^2) = 16\,|x|^2, \qquad g \in \partial B_\rho.$$

Proposition 3.7 presently gives

(3.15) $$\Sigma = \{g \in \partial B_\rho \,|\, x(g) = 0, |y(g)| = \frac{1}{4}\}.$$

Suppose we can show that there exists a constant $C > 0$ such that, given any $g \in \partial B_\rho$, one can find $\tilde{g} = \tilde{g}(g) \in \Sigma$ satisfying

(3.16) $$|x(g)| \geq C \, d(g, \tilde{g}).$$

Since $d(g, \tilde{g}) \geq d(g, \Sigma)$, in view of (3.14), (3.16) would imply (3.12), and thus complete the proof. Let then $g \in \partial B_\rho$. From (3.15) we find for $\tilde{g} \in \Sigma$

(3.17) $$d(g, \tilde{g})^4 = |x(g)|^4 + 16 \, |y(g) - y(\tilde{g})|^2.$$

We next distinguish two cases. If $y(g) = 0$, we obtain trivially from (3.13) $|x(g)| = 1$. We can thus take as $\tilde{g}$ any point in Σ since

$$d(g, \tilde{g}) \leq diam(B_\rho) = 2.$$

In this case, (3.16) holds trivially with $C = 1/2$. Suppose instead that $y(g) \neq 0$. If we choose $\tilde{g} = (0, y(\tilde{g}))$, with

$$y(\tilde{g}) = \frac{y(g)}{4|y(g)|},$$

then we see from (3.15) that $\tilde{g} \in \Sigma$. Moreover, it is easy to see that

$$|y(g) - y(\tilde{g})|^2 = \frac{1}{16} \, (1 - 4|y(g)|)^2.$$

From the latter equation and from (3.17) we conclude

(3.18) $$d(g, \tilde{g})^4 = |x(g)|^4 + (1 - 4\,|y(g)|)^2 = 2\,(1 - 4\,|y(g)|).$$

On the other hand, (3.13) gives

$$|x|^4 \geq (1 - 4\,|y|).$$

The latter inequality, and (3.18), imply that (3.16) holds with $C = 2^{-1/4}$. This completes the proof. □

Besides gauge balls, there exist large classes of manifolds $\mathcal{M}$ whose characteristic set is controlled. For the sake of simplicity, we discuss some of them in the context of the first Heisenberg group $\mathbb{H} = \mathbb{R}^3$, with its left invariant vector fields

(3.19) $$X_1 = \frac{\partial}{\partial x} + 2\,y\,\frac{\partial}{\partial t}, \qquad X_2 = \frac{\partial}{\partial y} - 2\,x\,\frac{\partial}{\partial t}$$

with respect to the non-commutative group law

(3.20) $$g \circ g' = (x, y, t) \circ (x', y', t') = (x + x', y + y', t + t' + 2(x'y - xy')).$$

The CC metric is equivalent to the gauge distance $\rho(g, g_o) = N(g^{-1} g_o)$, where we have denoted by

(3.21) $$N(g) = N(x, y, t) = [(|x|^2 + |y|^2)^2 + t^2]^{\frac{1}{4}}$$

the gauge defined in (2.14). The interested reader can easily generalize the considerations that follow to the higher Heisenberg groups $\mathbb{H}^n$. Using the results previously obtained, one can also extend some of the examples to groups of Heisenberg type.

EXAMPLE 3.11 (**Manifolds with cylindrical symmetry in the Heisenberg group**).

These manifolds have already been discussed in greater generality in the example of toroids. The novelty, here, is that we prove that their characteristic set is controlled. Let $\mathcal{M} \subset \mathbb{H}^1$ be a manifold whose defining function is given by $\phi(x, y, t) = t - f(|z|^4)$ where $f : [0, \infty) \to \mathbb{R}$ is a C^1 function, and $z = (x, y)$, $|z| = \sqrt{x^2 + y^2}$. Without loss of generality, we may assume $f(0) = 0$ (this situation can always be achieved by left-translation along the center). With $g = (x, y, t)$, $e = (0, 0, 0)$, we thus have

$$X_1\phi(g) = 4y - 8x|z|^2 f'(|z|^4), \quad X_2\phi(g) = -4x - 8y|z|^2 f'(|z|^4)$$

and

$$w(g)^2 = X_1\phi(g)^2 + X_2\phi(g)^2 = 16|z|^2 (1 + 4|z|^4 f'(|z|^4)^2).$$

In accordance with Proposition 3.7, we have $\Sigma = \{e\}$, therefore $d(g, \Sigma)^4 = |z|^4 + f(|z|^4)^2$. Since $f \in C^1$, and $f(0) = 0$, it is easy to see that

$$\lim_{g \to e} \frac{w(g)^4}{d(g, \Sigma)^4} = \lim_{g \to 0} \frac{4|z|^4(1 + 4|z|^4 f'(|z|^4)^2)^2}{|z|^4(1 + |z|^{-4} f(|z|^4)^2)} = 4 \lim_{\tau \to 0} \frac{1 + 4\tau f'(\tau)^2}{1 + \tau^{-1} f(\tau)^2} > 0.$$

We conclude that (3.12) holds. The same type of argument also applies to manifolds $\mathcal{M}$ whose defining function takes on the form $\phi(g) = t^2 - f(|z|^4)^2$.

EXAMPLE 3.12 (**The non-strongly isolated saddle in $\mathbb{H}^1$**).

D. Jerison in [**J81**] introduced the notion of strongly-isolated characteristic point in the Heisenberg group $\mathbb{H}^n$. Given a C^1 domain $\Omega \subset \mathbb{H}^n$, with defining function ϕ, a characteristic point $g_o \in \partial\Omega$ is called *strongly-isolated* if for some constant $C > 0$ one has for every $g \in \partial\Omega$

$$|X\phi(g)| \geq C\, d(g, g_o)\,.$$

This notion played an important role in [**J81**] in the study of the Dirichlet problem at characteristic points. In this example we produce a domain whose characteristic points are not strongly-isolated, according to the above definition, but whose characteristic set is nonetheless controlled. Consider the saddle in $\mathbb{H}^1$, defined by

$$\phi(g) = t - x^2 + y^2\,.$$

A computation gives

$$w(g) = |X\phi(g)| = 2\sqrt{2}\,|x - y|\,,$$

and therefore,

$$\Sigma = \{(\alpha, \alpha, 0) \mid \alpha \in \mathbb{R}\}\,.$$

Using calculus, one easily obtains that

$$d((x, y, x^2 - y^2), \Sigma) = \frac{1}{\sqrt{2}}\,|x - y|\,.$$

This estimate shows that (3.12) is satisfied, hence the characteristic set of the saddle is controlled. However, as it was observed in [**J81**], its characteristic points are not strongly-isolated.

EXAMPLE 3.13 (**Baouendi-Grushin vector fields and α-admissibility**).

In their cited paper [**MM02**] Monti and Morbidelli prove a trace theorem similar to Theorem 10.7 in the following two cases: a) The system $X = \{X_1, ..., X_m\}$ of C^∞ vector fields satisfies (1.4), the bounded domain Ω is C^∞, and it has empty characteristic set; b) In $\mathbb{R}^2$, the system $X = \{X_1, X_2\}$ consists of the Baouendi-Grushin vector fields

$$(3.22) \qquad X_1 = \frac{\partial}{\partial x}, \qquad X_2 = |x|^\alpha \frac{\partial}{\partial y}, \qquad \alpha > 0,$$

and the bounded domain Ω is α-*admissible*. As for case a), we have already observed that any non-characteristic domain is trivially controlled. Concerning b), we prove here that, at least in the range $0 < \alpha \leq 1$, the notion of α-admissibility is stronger than that of controlled characteristic set. Therefore, within this range, the class of domains introduced in Definition 1 in [**MM02**] have controlled characteristic set according to Definition 3.8. By left-translation along the y-axis, we can assume that the characteristic point occurs at the origin $(0,0)$. Definition 1 in [**MM02**] requires the existence of $f : (-\delta, \delta) \to \mathbb{R}$, such that $\partial\Omega$ can be locally described by the equation $y = f(x)$, with $f \in C^1$, $f(0) = 0$, and such that for some $M > 0$

$$(3.23) \qquad |f'(x)| \leq M |x|^\alpha, \qquad |x| < \delta.$$

We notice that (3.23) implies

$$(3.24) \qquad |f(x)| \leq M |x|^{\alpha+1}, \qquad |x| < \delta.$$

The function $\phi(x, y) = f(x) - y$ is a C^1 defining function for Ω near $(0,0)$. A simple computation gives

$$|X\phi(x,y)| = \sqrt{f'(x)^2 + |x|^{2\alpha}}, \qquad \text{on} \quad \partial\Omega,$$

therefore, thanks to (3.23) one has

$$(3.25) \qquad |x|^\alpha \leq |X\phi(x,y)| \leq C |x|^\alpha, \qquad \text{on} \quad \partial\Omega.$$

On the other hand, it is well-known [**J81**] that

$$d((x,y),(0,0)) \cong \left(|x|^{2(\alpha+1)} + y^2\right)^{\frac{1}{2(\alpha+1)}},$$

therefore, thanks to (3.24) we see that

$$(3.26) \qquad |x| \leq d((x,y),(0,0)) \leq C |x| \qquad \text{on} \quad \partial\Omega.$$

Combining (3.25) with (3.26), we see that near $(0,0)$ we have on $\partial\Omega$

$$\frac{|X\phi(x,y)|}{d((x,y), \Sigma)} \cong |x|^{\alpha-1}.$$

In particular, (3.12) is fulfilled, provided that $0 < \alpha \leq 1$. Some final comments are in order. The non-isotropic dilations attached to the vector fields (3.22) are $\delta_\lambda(x,y) = (\lambda x, \lambda^{\alpha+1} y)$. With respect to such dilations, a characteristic " convex cone" , see [**CG98**] and [**GV00**], is given by

$$\mathcal{C}_M = \{(x,y) \in \mathbb{R}^2 \mid y > M|x|^{\alpha+1}\},$$

where $M \geq 0$ measures the "aperture" of the cone. In this regard we see that condition (3.24) can be expressed by saying that near the characteristic point $\partial\Omega$ must stay below, or at most coincide with, the boundary of a cone $\mathcal{C}_M$.

EXAMPLE 3.14 (**A manifold whose characteristic set is not controlled**).

We close this chapter with an example of a manifold in $\mathbb{H}^1$ whose characteristic set is composed of a single point, and yet the condition in Definition 3.8 is not fulfilled. Consider the surface $\mathcal{M} \subset \mathbb{H}^1$ whose defining function is given by
$$\phi(g) = t - x^2 + y^2 + y^3.$$
A computation gives
$$X_1\phi(g) = -2x + 2y,$$
$$X_2\phi(g) = -2x + 2y + 3y^2.$$
Clearly, the only common solution to the above two equations is $x = y = 0$, hence $e = (0,0,0)$ is the only characteristic point. To see that $\Sigma = \{e\}$ it is not controlled, observe that for $g \in \mathcal{M}$ we have
$$w(g)^4 = (|X_1\phi(g)|^2 + |X_2\phi(g)|^2)^2 = (8(y-x)^2 + 12y^2(y-x) + 9y^4))^2$$
and
$$d(g,e)^4 = (x^2 + y^2)^2 + (x^2 - y^2 + y^3)^2.$$
Take a sequence $\alpha_n \to 0$. Considering points in $\mathcal{M}$ of the form $(\alpha_n, \alpha_n, -\alpha_n^3)$, we see that
$$\frac{w(\alpha_n, \alpha_n, -\alpha_n^3)^4}{d((\alpha_n, \alpha_n, -\alpha_n^3), e)^4} = \frac{81\alpha_n^8}{4\alpha_n^4 + \alpha_n^6} \longrightarrow 0 \quad \text{as } n \to \infty.$$
Hence, it is not possible to find a neighborhood V of $\Sigma = \{e\}$, and a constant $C > 0$, such that
$$w(g) \geq C\, d(g, \Sigma),$$
for all $g \in \mathcal{M} \cap V$.

CHAPTER 4

X-variation, X-perimeter and surface measure

The aim of this chapter is twofold. On the one hand, we recall the notions and some properties of the X-perimeter and of X-Caccioppoli sets. On the other hand, we introduce the concept of *perimeter measure*, which will play a pervasive role throughout the paper, and establish its connection with the X-perimeter, see Theorem 4.8. We begin by recalling the notion of $X-$variation introduced in [**CDG94**]. The latter is an intrinsic generalization of the original one due to De Giorgi [**DG54**], [**DG55**], [**DCP72**]. Two related definitions were independently set forth in [**BM95**], and in [**FSS96**]. In the latter paper, it was proved that the three definitions in [**CDG94**], [**BM95**], [**FSS96**], are in fact equivalent. A generalization of the notion of variation to a metric space with a doubling measure and a Poincaré inequality has been given in [**Mi00**].

DEFINITION 4.1. Let $\Omega \subset \mathbb{R}^n$ be an open set, and $u \in L^1_{loc}(\Omega)$. The $X-$*variation* of u in Ω is defined as follows

$$Var_X(u;\Omega) = \sup_{\zeta \in \mathcal{F}(\Omega)} \int_\Omega u \sum_{j=1}^m X_j^* \zeta_j \, dx \, ,$$

where

$$\mathcal{F}(\Omega) = \left\{ \zeta = (\zeta_1, ..., \zeta_m) \in C_o^1(\Omega)^m \mid ||\zeta||_\infty = \sup_{x \in \Omega} \left(\sum_{j=1}^m \zeta_j(x)^2 \right)^{1/2} \leq 1 \right\}.$$

A function $u \in L^1(\Omega)$ is called of *bounded $X-$variation* if $Var_X(u;\Omega) < \infty$. In such case, we write $u \in BV_X(\Omega)$, and the collection of all such functions becomes a Banach space when endowed with the norm

$$||u||_{BV_X(\Omega)} = ||u||_{L^1(\Omega)} + Var_X(u;\Omega).$$

The notation $BV_{X,loc}(\Omega)$ indicates the collection of functions $u \in L^1_{loc}(\Omega)$, such that $u \in BV_X(\omega)$, for every $\omega \subset\subset \Omega$. A basic source for the properties of the space BV_X is [**GN96**], where also the existence of minimal surfaces was established. In the same paper we posed the question of their regularity properties. For some very interesting progress in this direction one should consult the papers [**FSS01**], [**FSS03(I)**], [**FSS04**], [**M04**].

4.1. The structure of functions in $BV_{X,loc}$

An important consequence of Definition 4.1, and of the Riesz representation theorem, is the following structure theorem for $BV_{X,loc}$ functions. Hereafter, we denote by $\mathcal{R}(\Omega)$ the space of Radon measures on Ω, and by $\mathcal{R}(\Omega)^m$ that of m-vector valued ones.

THEOREM 4.2. *Let $\Omega \subset \mathbb{R}^n$ be an open set, and $u \in BV_{X,loc}(\Omega)$. There exists $\mu_u \in \mathcal{R}(\Omega)$, and a μ_u-measurable function $\sigma^u = (\sigma_1^u, ..., \sigma_m^u) : \Omega \to \mathbb{R}^m$, such that*

(i) $\quad |\sigma^u(x)| = 1 \quad\quad \mu_u - a.e.\ x \in \Omega;$

(ii) $\displaystyle \int_\Omega u \sum_{j=1}^m X_j^* \zeta_j\, dx = - \int_\Omega \sum_{j=1}^m \zeta_j\, \sigma_j^u\, d\mu_u(x)$

for every $\zeta = (\zeta_1, ..., \zeta_m) \in C_o^1(\Omega; \mathbb{R}^m)$.

Conforming to a well established tradition, for $u \in BV_{X,loc}(\Omega)$ we introduce the notation

(4.1) $\quad\quad ||Xu|| = \mu_u, \quad\quad [Xu] = ||Xu|| \lfloor \sigma_u.$

where μ_u is the *variation measure* in Theorem 4.2. Notice that $d[Xu] \in \mathcal{R}(\Omega)^m$. Equation (ii) can thus be written

(4.2) $\displaystyle \int_\Omega u \sum_{j=1}^m X_j^* \zeta_j\, dx = -\int_\Omega <\zeta, \sigma^u>\, d||Xu|| = -\int_\Omega <\zeta, d[Xu]>,$

for every $\zeta \in C_o^1(\Omega; \mathbb{R}^m)$, where we have indicated with $<\cdot,\cdot>$ the inner product in $\mathbb{R}^m$.

4.2. X-Caccioppoli sets

The following definitions are respectively taken from [**CDG94**] and [**GN96**].

DEFINITION 4.3. *Let $\Omega \subset \mathbb{R}^n$ be an open set. Given a measurable set $E \subset \mathbb{R}^n$, the X-perimeter of E relative to Ω is defined by*

$$P_X(E;\Omega) = Var_X(\chi_E;\Omega).$$

DEFINITION 4.4. *Let $\Omega \subset \mathbb{R}^n$ be an open set. A measurable set $E \subset \mathbb{R}^n$ is called a X-Caccioppoli set in Ω if*

$$P_X(E;\omega) < \infty,$$

for every $\omega \subset\subset \Omega$. Equivalently, $E \subset \mathbb{R}^n$ is a X-Caccioppoli set in Ω if

$$\chi_E \in BV_{X,loc}(\Omega).$$

When $E \subset \mathbb{R}^n$ is a X-Caccioppoli set in Ω, taking $u = \chi_E$ in Theorem 4.2, we will write

(4.3) $\quad ||\partial_X E|| = \mu_{\chi_E}, \quad\quad \nu_X^E = \sigma^{\chi_E}, \quad\quad [\partial_X E] = ||\partial_X E|| \lfloor \nu_X^E,$

and respectively call $||\partial_X E||$ the X-*perimeter measure*, and ν_X^E the *generalized X-outer normal* to E. If E is a X-Caccioppoli set in $\mathbb{R}^n$, and $\Omega \subset \mathbb{R}^n$ is an open set, with the notation in (4.3) we have

$$||\partial_X E||(\Omega) = P_X(E;\Omega).$$

Let $\zeta \in C_o^1(\mathbb{R}^n; \mathbb{R}^m)$, then for any measurable set $E \subset \mathbb{R}^n$ one has

$$\int_E \sum_{j=1}^m X_j^* \zeta_j \, dx + \int_{E^c} \sum_{j=1}^m X_j^* \zeta_j \, dx = \int_{\mathbb{R}^n} \sum_{j=1}^m X_j^* \zeta_j \, dx = 0 \ .$$

This implies

(4.4) $\qquad P_X(E; \mathbb{R}^n) = P_X(E^c; \mathbb{R}^n) \ , \qquad \nu_X^E = - \nu_X^{E^c} \ .$

Because of its relevance, we will restate Theorem 4.2 for Caccioppoli sets, using the notations in (4.3).

THEOREM 4.5. *Given an open set $\Omega \subset \mathbb{R}^n$, let $E \subset \mathbb{R}^n$ be a X-Caccioppoli set in Ω. There exists a $||\partial_X E||$-measurable function $\nu_X^E : \Omega \to \mathbb{R}^m$, such that*

$$|\nu_X^E(x)| = 1 \qquad \text{for } ||\partial_X E|| - a.e. \, x \in \Omega \ ,$$

and for which one has for every $\zeta \in C_o^1(\Omega; \mathbb{R}^m)$

$$\int_E \sum_{j=1}^m X_j^* \zeta_j \, dx = - \int_\Omega <\zeta, \nu_X^E> d||\partial_X E|| = - \int_\Omega <\zeta, d[\partial_X E]> \ .$$

For our purposes, the following property of the X-perimeter is important, see Remark 3.2 in [**CDG94**]. Let $E \subset \mathbb{R}^n$ be a C^1 domain, with outer unit normal ν, and assume that $H_{n-1}(\Omega \cap \partial E) < \infty$. If $\zeta \in C_o^1(\Omega; \mathbb{R}^m)$, we have

$$\int_E \sum_{j=1}^m X_j^* \zeta_j \, dx = - \int_{\partial E \cap \Omega} \sum_{j=1}^m \zeta_j <X_j, \nu> dH_{n-1} \ .$$

From this observation, and from Theorem 4.5, we conclude the following result.

PROPOSITION 4.6. *Let $E \subset \mathbb{R}^n$ be a C^1 domain, and denote by ν its outward pointing unit normal. If for any given compact set $K \subset \mathbb{R}^n$, one has $H_{n-1}(\partial E \cap K) < \infty$, then for every open set $\Omega \subset \mathbb{R}^n$, and any $\zeta \in C_o^1(\Omega; \mathbb{R}^m)$, one has*

$$\int_\Omega <\zeta, \nu_X^E> d||\partial_X E|| = \int_{\partial E \cap \Omega} <\zeta, \vec{X}_\nu> dH_{n-1} \ ,$$

where we have let

(4.5) $\qquad \vec{X}_\nu = \left(<X_1, \nu>, ..., <X_m, \nu> \right) \ .$

Moreover,

$$d||\partial_X E|| = |\vec{X}_\nu| \, d(H_{n-1} \lfloor \partial E) = \left\{ \sum_{j=1}^m <X_j, \nu>^2 \right\}^{1/2} d(H_{n-1} \lfloor \partial E) \ ,$$

and one has

(4.6) $\qquad P_X(E; \Omega) = ||\partial_X E||(\Omega) = \int_{\partial E \cap \Omega} |\vec{X}_\nu| \, dH_{n-1} \ .$

We observe that, when $x_o \in \partial E \cap \Omega$ is a characteristic point of E, then we obtain from (4.5)

$$\vec{X}_\nu(x_o) = (0, ..., 0) \in \mathbb{R}^m \ .$$

However, we recall that Theorem 3.2 states that, when Ω is C^∞, the characteristic set $\Sigma_{E,X}$ has H_{n-1} measure zero.

4.3. X-perimeter and the perimeter measure

We begin with a definition that plays an important role in this work.

DEFINITION 4.7. Let $\Omega \subset \mathbb{R}^n$ be an open set for which there exists $\phi \in C^1(\mathbb{R}^n)$ such that
$$\Omega = \{x \in \mathbb{R}^n \mid \phi(x) < 0\} .$$
We define the *perimeter measure* μ (associated with the system X and with Ω) as follows

$$(4.7) \quad \mu(E) \stackrel{def}{=} \int_{E \cap \partial \Omega} |X\phi(x)| \, dH_{n-1}(x) , \qquad E \text{ is a Borel subset of } \mathbb{R}^n .$$

Clearly, μ is supported on $\partial \Omega$.

The following result provides a geometric interpretation of the measure μ, by showing that it charges the intersection of a CC ball with the boundary of a C^1 domain Ω proportionally to the X−perimeter of Ω with respect to the ball. In this situation, μ is therefore nothing but the X-perimeter in disguise, thereby justifying its name.

THEOREM 4.8. *Let $\Omega \subset \mathbb{R}^n$ be a C^1 domain, with defining function ϕ satisfying (3.4). For every $x \in \partial \Omega$, and every $r > 0$, we have*
$$\alpha \, \mu(B(x,r)) \leq P_X(\Omega; B(x,r)) \leq \beta \, \mu(B(x,r)).$$

Proof. If we denote with ν the outward unit normal to $\partial \Omega$, we then have $\nu = \nabla \phi / |\nabla \phi|$. Proposition 4.6 gives

$$(4.8) \quad P_X(\Omega; B(x,r)) = \int_{\partial \Omega \cap B(x,r)} \left\{ \sum_{j=1}^m <X_j, \nu>^2 \right\}^{1/2} dH_{n-1}$$
$$= \int_{\partial \Omega \cap B(x,r)} \frac{|X\phi|}{|\nabla \phi|} \, dH_{n-1} .$$

The conclusion now follows from (3.4), and from (4.7). □

CHAPTER 5

Geometric estimates from above on CC balls for the perimeter measure

This chapter, and the next one, have a twofold purpose. On one hand, they serve as motivation for the rest of the paper by showing that the Ahlfors type assumptions in our main results hold generically in the geometric setting of Carnot groups for the X-perimeter measure. At the same time, they provide a solid foundation for the development of the theory. The results in this chapter, and those in Chapter 6, have also important applications in partial differential equations, especially in the study of the Dirichlet and Neumann problems, see [**CGN04**], [**CGN02**], [**LU02**], in the theory of free boundaries [**DGS02**], [**DGP05**], and in the development of geometric measure theory in CC spaces [**DGN04(II)**].

5.1. A fundamental estimate

Let $\mathbf{G}$ be a Carnot group, with Lie algebra $\mathfrak{g}$, and topological dimension N. We consider the $(N-1)$−dimensional Hausdorff measure H_{N-1} in $\mathfrak{g}$ constructed with the standard Euclidean distance. For a given open set $D \subset \mathfrak{g}$ we will denote with $d\sigma_\mathfrak{g}$ the measure on ∂D defined by

(5.1) $$\sigma_\mathfrak{g} = H_{N-1} \lfloor \partial D.$$

If D is a Lipschitz domain, then $d\sigma_\mathfrak{g}$ is just the ordinary surface measure on ∂D. Slightly abusing the notation, we will also indicate with H_{N-1} the $(N-1)$−dimensional Hausdorff measure in $\mathbf{G}$ constructed with the Riemannian distance. However, no confusion will arise since we will ordinarily write $dH_{N-1}(\xi)$, or $dH_{N-1}(g)$, depending on whether the integral in question is performed with respect to the variable $\xi \in \mathfrak{g}$, or $g \in \mathbf{G}$. If $\Omega \subset \mathbf{G}$ is an open set, and if $D = \exp^{-1}(\Omega)$, then we define a measure on $\partial \Omega$ via the formula

(5.2) $\quad \sigma(E) = \sigma_\mathfrak{g}(\exp^{-1}(E)), \quad$ for every measurable $\quad E \subset \partial \Omega,$

where $\sigma_\mathfrak{g}$ is given by (5.1). Clearly, $E \subset \partial \Omega$ is called measurable, if $\exp^{-1}(E)$ is a measurable subset of ∂D with respect to $\sigma_\mathfrak{g}$. Our main present objective is providing a significant example of an upper s-Ahlfors measure μ on $\mathbf{G}$, i.e., a non-negative Borel measure with the property

(5.3) $$\mu(B(g,r)) \leq M \frac{|B(g,r)|}{r^s}.$$

Throughout this chapter, we will in fact be solely concerned with the geometrically important case $s = 1$. We begin by recalling a basic notion introduced in [**CG05**].

DEFINITION 5.1. Let G be a Carnot group of step r. Given a bounded open set
$$\Omega = \{g \in G \mid \phi(g) < 0\},$$
where $\phi \in C^1(G)$, we define the "type" of $g_o \in \partial\Omega$ to be the smallest $j = 1, ..., r$ such that there exists $s = 1, ..., m_j$ for which
$$X_{j,s}\phi(g_o) \neq 0.$$
Such integer will be denoted by $type(g_o)$. If for every $g_o \in \partial\Omega$ we have $type(g_o) \leq k$, then we say that Ω has type $\leq k$. In particular, Ω is of type 1 if and only if it has empty characteristic set.

REMARK 5.2. If G is of step 2, then in view of Lemma 6.7 below any bounded C^1 domain $\Omega \subset G$ is of type ≤ 2, so that Definition 5.1 imposes in this case no restriction on the characteristic set.

It is easy to generalize the notion of type ≤ 2 to a C^∞ system $X = \{X_1, ..., X_m\}$ satisfying (1.4). Given a bounded C^1 domain $\Omega = \{x \in \mathbb{R}^n \mid \phi(x) < 0\}$, we say that Ω is of type ≤ 2 if either Ω is non-characteristic, or for every characteristic point $x_o \in \partial\Omega$ one has
$$[X_i, X_j]\phi(x_o) \neq 0,$$
for some $i, j \in \{1, ..., m\}$.

THEOREM 5.3 ([**DGN98**], [**CGN98**], [**CGN02**], [**CG05**]). *Let G be a Carnot group of arbitrary step, having homogeneous dimension Q. Consider a bounded domain of type ≤ 2*
$$\Omega = \{g \in G \mid \phi(g) < 0\},$$
where $\phi \in C^{1,1}(G)$ is a defining function for Ω. There exist $M, R_o > 0$, depending on G and Ω, such that for every $g_o \in \partial\Omega$, and $0 < r \leq R_o$, one has

(5.4) $$\left(\sup_{B(g_o,r) \cap \partial\Omega} |X\phi|\right) \sigma(B(g_o, r) \cap \partial\Omega) \leq M \, r^{Q-1}.$$

REMARK 5.4. As we discussed in the introduction, the assumption of type ≤ 2 in Theorem 5.3 is best possible. In [**CG05**] the authors have constructed an example of a C^∞ domain Ω of type 3 in the 4-dimensional Engel group such that the estimate (5.4) fails. Furthermore, for the same domain also the perimeter measure $P_X(\Omega; \cdot)$ fails to be upper 1-Ahlfors, and therefore the type condition is optimal for Theorem 5.6 below as well.

We mentioned that, in the special setting of the Heisenberg group $\mathbb{H}^n$, and for C^2 domains, versions of Theorem 5.3 were independently formulated by C. Romero and by M. Mekias, in their respective Ph.D. Dissertations [**Ro91**] and [**Me93**]. A complete proof for the Heisenberg group first appeared in [**DGN98**]. With minor modifications, such result was extended to all Carnot groups of step 2 in [**CGN02**]. The proof of the full Theorem 5.3 will appear in the paper [**CG05**]. In fact, in [**CG05**], Theorem 5.3 is one of the main steps in a chain of arguments which, ultimately exploiting the Rothschild-Stein lifting theorem [**RS76**], establishes the following analogous result for any system X of Hörmander type.

THEOREM 5.5 ([**CGN04**]). *Let X be a system of free C^∞ vector fields in $\mathbb{R}^n$ satisfying (1.4). Consider a bounded open set*

$$\Omega = \{x \in \mathbb{R}^n \mid \phi(x) < 0\},$$

where $\phi \in C^{1,1}(\mathbb{R}^n)$ is a defining function for Ω. If Ω is of type ≤ 2 there exist $M, R_o > 0$, depending on X and on Ω, such that for every $x_o \in \partial\Omega$, and $0 < r \leq R_o$, one has

$$\left(\sup_{B(x_o,r) \cap \partial\Omega} |X\phi| \right) \sigma(B(x_o,r) \cap \partial\Omega) \leq M \frac{|B(x_o,r)|}{r},$$

where we have let σ denote the standard surface measure on $\partial\Omega$.

We have already pointed out that the geometric estimates in Theorems 5.3, 5.5 play a fundamental role in several questions which range from boundary value problems, to geometric measure theory. In this paper, they will be primarily used in connection with trace inequalities. We observe explicitly that Theorems 5.3, 5.5 provide a precise quantitative information on how bad surface measure $d\sigma$ can be at characteristic points. For instance, in the case of a Carnot group of step 2 it was shown in [**DGN98**] and [**CGN02**] that one has

$$\sigma(B(g_o,r) \cap \partial\Omega) \leq M \frac{|B(g_o,r)|}{r^2},$$

and that such estimate is sharp at characteristic points, in the sense that it cannot be improved to

$$\sigma(B(g_o,r) \cap \partial\Omega) \leq M \frac{|B(g_o,r)|}{r^s},$$

for some $1 \leq s < 2$. Away from the characteristic set one expects surface measure to be well-behaved. This intuition has been proved correct by Monti and Morbidelli in their cited paper [**MM02**]. They show that for a bounded C^∞ domain Ω having empty characteristic set, the following interesting estimate holds

$$M^{-1} \frac{|B(x_o,r)|}{r} \leq \sigma(B(x_o,r) \cap \partial\Omega) \leq M \frac{|B(x_o,r)|}{r} \qquad x_o \in \partial\Omega, \quad 0 < r \leq R_o.$$

5.2. The X-perimeter of a $C^{1,1}$ domain is an upper 1-Ahlfors measure

These observations naturally lead to the question of what is the appropriate replacement for surface measure in the trace inequalities. The answer is contained in the following important consequence of the above theorems.

THEOREM 5.6. *Let $\Omega \subset \mathbb{R}^n$ be as in Theorem 5.5. There exist $M, R_o > 0$, depending on X and Ω, such that the perimeter measure μ on Ω, introduced in Definition 4.7, satisfy the estimate*

$$\mu(B(x_o,r)) \leq M \frac{|B(x_o,r)|}{r},$$

for every $x_o \in \partial\Omega$, and any $0 < r \leq R_o$. This estimate, along with Theorem 4.8, imply that μ and $P_X(\Omega; \cdot)$ are upper 1-Ahlfors measures.

Proof. Using Theorem 5.3 one finds

$$\begin{aligned}
\mu(B(g_o,r)) &= \int_{B(g_o,r)\cap \partial\Omega} |X\phi(g)|\, d\sigma \\
&\leq \left(\sup_{B(g_o,r)\cap\partial\Omega} |X\phi| \right) \sigma(B(g_o,r)\cap\partial\Omega) \\
&\leq M \frac{|B(g_o,r)|}{r}.
\end{aligned}$$

$\square$

REMARK 5.7. One should compare the estimate in Theorem 5.6 with that in the right-hand side of (1.1).

CHAPTER 6

Geometric estimates from below on CC balls for the perimeter measure

We now turn to the study of the bound from below for the perimeter measure introduced in Definition 4.7. Due to the presence of characteristic points on the boundary, this constitutes a delicate endeavor. Our main result is contained in the following Theorem 6.1. We emphasize that, in the statement of the latter, no assumption is made on the characteristic set of the domain Ω.

THEOREM 6.1. *Let G be a Carnot group of step 2, with Lie algebra $\mathfrak{g} = V_1 \oplus V_2$, and consider a bounded $C^{1,1}$ domain $\Omega \subset G$. There exist positive constants M, R_o, depending on Ω, such that if μ denotes the perimeter measure associated with an orthonormal basis of the first layer V_1, one has for every $g_o \in \partial\Omega$, and any $0 < r < R_o$,*

$$\mu(B(g_o, r)) \geq M^{-1} \frac{|B(g_o, r)|}{r}.$$

Combining this estimate with Theorem 4.8, we conclude that μ and $P_X(\Omega; \cdot)$ are lower 1-Ahlfors measures.

We present two very different proofs of Theorem 6.1. The former, based on powerful tools from geometric measure theory, is relatively short and is, in essence, contained in the proofs of Theorems 1.7 and 1.8 in [**DGN98**]. The latter, instead, is a direct proof which is based on calculus. From the point of view of the present work, the merit of the direct proof is twofold. On one hand, the main estimates in it are of independent interest and might prove useful in other situations. On the other hand, it brings to light all the difficulties which are connected with the characteristic set, whereas such aspects are completely hidden in the geometric measure theory proof. The underlying reason for this is that, since under the regularity assumptions in which we work the perimeter measure is mutually absolutely continuous with respect to the Hausdorff measure $\mathcal{H}^{Q-1}$ (for the latter see Definition 7.9), and since the characteristic set has the property $\mathcal{H}^{Q-1}(\Sigma) = 0$, see the cited papers [**B00**], [**M03(I)**], and [**M03(II)**], then the perimeter measure does not see Σ.

Before turning to the proofs we pause to list a basic consequence of Theorems 4.8, 5.3, and 6.1.

THEOREM 6.2. *In a Carnot group G of step 2, let $\Omega \subset G$ be a bounded $C^{1,1}$ domain. There exist constants $M = M(G, \Omega) > 0$, $R_o = R_o(G, \Omega) > 0$, such that for every $g_o \in \partial\Omega$, and any $0 < r < R_o$ one has*

$$M^{-1} \frac{|B(g_o, r)|}{r} \leq P_X(\Omega; B(g_o, r)) \leq M \frac{|B(g_o, r)|}{r},$$

i.e., $P_X(\Omega; \cdot)$ is a 1-Ahlfors measure. In particular, $P_X(\Omega; \cdot)$ is doubling, i.e., there exists $C = C(\boldsymbol{G}, \Omega) > 0$ such that

$$P_X(\Omega; B(g_o, 2r)) \leq C \, P_X(\Omega; B(g_o, r)) \qquad 0 < r < R_o/2 \, .$$

6.1. The relative isoperimetric inequality and Theorem 6.1

We will need the following result, which is a special case of Theorem 1.18 in [**GN96**].

THEOREM 6.3. *Let $\boldsymbol{G}$ be a Carnot group, and let $S \subset \boldsymbol{G}$ be a PS (Poincaré-Sobolev) domain. There exists a constant $C_{iso} > 0$, depending only on G, and on the PS parameters of S, such that for any set $\Omega \subset \boldsymbol{G}$ with locally finite X-perimeter one has*

$$\min\left(|\Omega \cap S|, |\Omega^c \cap S|\right)^{\frac{Q-1}{Q}} \leq C_{iso} \, diam(S) \, |S|^{-\frac{1}{Q}} \, P_X(\Omega; S) \, .$$

Given an open set $\Omega \subset \boldsymbol{G}$, we say that Ω admits an interior corkscrew at $g_o \in \partial\Omega$ if for some $K, R_o > 0$, for any $0 < r < R_o$ one can find $A_r(g_o) \in \Omega$ (a corkscrew) such that

$$\frac{r}{K} < d(A_r(g_o), g_o) \leq r \,, \qquad dist(A_r(g_o), \partial\Omega) > \frac{r}{K} \, .$$

If in the above inequalities the same K and R_o can be chosen for every $g_o \in \partial\Omega$, then we say that Ω has the uniform interior corkscrew condition. Finally, Ω is said to satisfy the *uniform corkscrew condition* if both Ω and Ω^c fulfill the uniform interior corkscrew condition. We will need the following result from the statement of Theorem 2.24 in [**CG98**].

THEOREM 6.4. *Let $\boldsymbol{G}$ be a Carnot group of step $r = 2$, then every bounded $C^{1,1}$ domain $\Omega \subset \boldsymbol{G}$ satisfies the uniform corkscrew condition.*

Geometric measure theory proof of Theorem 6.1. According to (I) in Theorem 1.15 in [**GN96**] every CC ball $B(g, r)$ is a PS domain in $\boldsymbol{G}$ (the same result was proved independently in [**FGW94**]). Since a $C^{1,1}$ domain $\Omega \subset \boldsymbol{G}$ is, in particular, a set with locally finite X-perimeter, taking $S = B(g_o, r)$ in Theorem 6.3 we obtain for every $g_o \in \partial\Omega$ and every $r > 0$

$$\min\left(|\Omega \cap B(g_o, r)|, |\Omega^c \cap B(g_o, r)|\right)^{\frac{Q-1}{Q}} \leq C_{iso} \frac{diam(B(g_o, r))}{|B(g_o, r)|^{\frac{1}{Q}}} P_X(\Omega; B(g_o, r)) \, .$$

Since thanks to Theorem 6.4, Ω possesses the uniform corkscrew condition, there exist $K, R_o > 0$, depending on $\boldsymbol{G}$ and on the $C^{1,1}$-character of Ω, such that for every $g_o \in \partial\Omega$ and $0 < r < R_o$ one has

$$K \, \frac{|B(g_o, r)|}{r} \leq \min\left(|\Omega \cap B(g_o, r)|, |\Omega^c \cap B(g_o, r)|\right)^{\frac{Q-1}{Q}} ,$$

where we have used that $diam(B(g_o, r)) \leq 2r$. Combining the latter two inequalities, we obtain

$$C \, \frac{|B(g_o, r)|}{r} \leq P_X(\Omega; B(g_o, r)) \, ,$$

for some constant $C = C(\boldsymbol{G}, \Omega) > 0$. In view of Theorem 4.8 this concludes the proof. $\square$

We stress that in [**GN96**] Theorem 6.3 was shown to hold for a general CC space associated with a system of Lipschitz vector fields satisfying a doubling condition such as (1.22) and a weaker form of the Poincaré inequality in Theorem 7.1. Analyzing the above proof we thus infer that the following more general result actually holds.

THEOREM 6.5. *In a Carnot group $\boldsymbol{G}$ (or more in general in a CC space with a doubling and a Poincaré inequality) let $\Omega \subset \boldsymbol{G}$ be a bounded open set with locally finite X-perimeter and satisfying the uniform corkscrew condition. There exist constants $M = M(\boldsymbol{G}, \Omega) > 0$, $R_o = R_o(\boldsymbol{G}, \Omega) > 0$, such that for every $g_o \in \partial\Omega$, and any $0 < r < R_o$ one has*

$$M^{-1} \frac{|B(g_o, r)|}{r} \leq P_X(\Omega; B(g_o, r)),$$

i.e., $P_X(\Omega; \cdot)$ is a lower 1-Ahlfors measure.

As we have already said, in the above proof the characteristic set seems to play no role. To better elucidate the role of such set, we next present, under the hypothesis that Ω be C^2, instead of $C^{1,1}$, a direct proof of Theorem 6.1. Such proof avoids altogether the use of the powerful tools from geometric measure theory, such as the relative isoperimetric inequality in Theorem 6.3. and has the advantage of highlighting the delicate role of the characteristic set in obtaining bounds from below for the perimeter measure. We begin with some preliminary steps.

6.2. A basic geometric lemma

In what follows, for a given function $f : \boldsymbol{G} \to \mathbb{R}$, we let $|Yf| = (\sum_{s=1}^{k}(Y_s f)^2)^{1/2}$, where the vector fields Y_s are like in (2.3).

LEMMA 6.6. *Given a Carnot group $\boldsymbol{G}$ of step 2, let $\Omega \subset \boldsymbol{G}$ be a C^2 bounded domain with defining function ϕ. There exist constants $R_1, C > 0$, depending on $\boldsymbol{G}$ and Ω, such that for all $g_o \in \partial\Omega$, and $0 < r \leq R_1$, the following estimate holds*

$$(6.1) \qquad C \frac{r^{Q-1}}{\eta(g_o, r)} \leq \sigma(B(g_o, r) \cap \partial\Omega) \leq C^{-1} \frac{r^{Q-1}}{\eta(g_o, r)},$$

where Q is the homogeneous dimension of $\boldsymbol{G}$, and

$$(6.2) \qquad \eta(g_o, r) = \frac{\max\{|X\phi(g_o)|, r\,|Y\phi(g_o)|\}}{|\nabla \phi(g_o)|}.$$

Proof. We divide the proof in two steps. First, we obtain an estimate of $\sigma_{\mathfrak{g}}(\Pi \cap Box(r))$, where

$$\Pi = \{(x, y) \in \mathfrak{g} \mid A_1 x_1 + \cdots + A_m x_m + B_1 y_1 + \cdots + B_k y_k = 0\}$$

is an arbitrary hyper-plane in $\mathfrak{g}$ passing through the origin, and $Box(r)$ is the non-isotropic box defined by

$$Box(r) = \{\xi = \xi_1 + \xi_2 \in \mathfrak{g} \mid |\xi_1| < r, |\xi_2| < r^2\}.$$

We notice explicitly that this definition of $Box(r)$ coincides with that in (1.27) for a general system of vector fields of Hörmander type. In the second step, we show that the sought for estimate of $\sigma(\Delta(g_o, r))$ can be derived from that of $\sigma_{\mathfrak{g}}(\Pi \cap Box(r))$, found in step one.

Step 1. Let in $\mathfrak{g} = V_1 \oplus V_2$ be the Lie algebra of $\mathbf{G}$. We observe that the set $Box(r)$ is invariant with respect to the action of the orthogonal groups $O(m)$ and $O(k)$ on V_1 and V_2, respectively. Furthermore, the same is true for the measure H_{N-1}. This is not completely obvious, but a moment's thought should convince the reader of the validity of this fact. Performing an orthogonal transformation inside each layer V_i, we can thus assume without loss of generality that the defining equation for Π is given by

$$\pi(X, Y) = A\, x_1 + B\, y_1, \tag{6.3}$$

where $A = \left(\sum_{j=1}^{m} A_i^2\right)^{\frac{1}{2}}$ and $B = \left(\sum_{j=1}^{k} B_j^2\right)^{\frac{1}{2}}$. With Π defined by (6.3), we obviously have

$$Box(r) \cap \Pi = \{(x,y) \in \mathbb{R}^{m+k} \mid Ax_1 + By_1 = 0,\ |x| \leq r,\ |y| \leq r^2\}.$$

Moreover, by an isometric linear map, we can transform the set

$$\{(x,y) \in \mathbb{R}^{m+k} \mid Ax_1 + By_1 = 0,\ |x| \leq r,\ |y| \leq r^2\}$$

into

$$(-r, r)^{m-1} \times (-r^2, r^2)^{k-1} \times S,$$

where

$$S = \{(s_1, s_2) \in \mathbb{R}^2 \mid As_1 + Bs_2 = 0,\ |s_i| \leq r^i,\ i = 1, 2\}.$$

These considerations allow to conclude

$$\sigma_{\mathfrak{g}}(\Pi \cap Box(r)) = 2^{N-2}\, r^{Q-3}\, |S|, \tag{6.4}$$

where $|S|$ denotes the length of the segment S. By considering the two cases $A \leq B\,r$, or $A > B\,r$, it is easy to see that

$$|S| = \frac{2\, r^2\, \sqrt{A^2 + B^2}}{max(A,\ B\,r)}. \tag{6.5}$$

Hence, using (6.5) in (6.4), and recalling the definitions of A and B, we finally obtain

$$\sigma_{\mathfrak{g}}(\Pi \cap Box(r)) = 2^{N-1}\, \frac{r^{Q-1}}{\eta(r)}, \tag{6.6}$$

where

$$\eta(r) = \frac{max\left(\left(\sum_{j=1}^{m} A_j^2\right)^{\frac{1}{2}}, \left(\sum_{l=1}^{k} B_l^2\right)^{\frac{1}{2}} r\right)}{\sqrt{\sum_{j=1}^{m} A_j^2 + \sum_{l=1}^{k} B_l^2}}. \tag{6.7}$$

Step 2. In this second step, we begin by observing that in view of (5.2), if $g_o = \exp(\xi_o)$, and $\Omega = \exp(D)$, then

$$\sigma(\Delta(g_o, r)) = \sigma_{\mathfrak{g}}(\exp^{-1}(\Delta(g_o, r)) = \sigma_{\mathfrak{g}}(\partial D \cap B_{\mathfrak{g}}(\xi_o, r)).$$

Using Taylor's formula, and the compactness of ∂D, we can approximate $\sigma_{\mathfrak{g}}(\partial D \cap B_{\mathfrak{g}}(\xi_o, r))$ with $\sigma_{\mathfrak{g}}(T_{\xi_o}(\partial D) \cap B_{\mathfrak{g}}(\xi_o, r))$, i.e., there exists a constant $C > 0$, depending only on $\boldsymbol{G}$, and Ω, such that

$$(6.8) \qquad C \leq \frac{\sigma_{\mathfrak{g}}(\partial D \cap B_{\mathfrak{g}}(\xi_o, r))}{\sigma_{\mathfrak{g}}(T_{\xi_o}(\partial D) \cap B_{\mathfrak{g}}(\xi_o, r))} \leq C^{-1}.$$

We next use the ball-box Theorem 1.17 to find $C = C(\boldsymbol{G}) > 0$, such that

$$(6.9) \qquad \theta_{\xi_o}(Box(Cr)) \subset B_{\mathfrak{g}}(\xi_o, r) \subset \theta_{\xi_o}(Box(C^{-1}r)),$$

where θ_{ξ_o} is defined in (2.8). Using the inclusions (6.9), we conclude that

$$(6.10) \qquad \sigma_{\mathfrak{g}}(T_{\xi_o}(\partial D) \cap \theta_{\xi_o}(Box(Cr)))$$
$$\leq \sigma_{\mathfrak{g}}(T_{\xi_o}(\partial D) \cap B_{\mathfrak{g}}(\xi_o, r))$$
$$\leq \sigma_{\mathfrak{g}}(T_{\xi_o}(\partial D) \cap \theta_{\xi_o}(Box(C^{-1}r))).$$

From now on, we concentrate on estimating one of the two quantities in the left- and right-hand side of (6.10). For simplicity, we neglect the immaterial factor C, and consider

$$\sigma_{\mathfrak{g}}(T_{\xi_o}(\partial D) \cap \theta_{\xi_o}(Box(r))) .$$

The map θ_{ξ_o} is one-to-one, therefore we have

$$\sigma_{\mathfrak{g}}(T_{\xi_o}(\partial D) \cap \theta_{\xi_o}(Box(r))) = \sigma_{\mathfrak{g}}\left(\theta_{\xi_o}\left(\theta_{\xi_o}^{-1}(T_{\xi_o}(\partial D)) \cap Box(r)\right)\right).$$

We now notice that in a Carnot group of step 2 the maps $\theta_{\xi_o}, \theta_{\xi_o^{-1}} : \mathfrak{g} \to \mathfrak{g}$ are affine, see (2.20), therefore globally Lipschitz with respect to the Euclidean metric in $\mathfrak{g}$. By Theorem 1 in sec. 2.4.1 in [**EG92**], we infer

$$(6.11) \qquad \sigma_{\mathfrak{g}}\left(\theta_{\xi_o}\left(\theta_{\xi_o^{-1}}(T_{\xi_o}(\partial D)) \cap Box(r)\right)\right)$$
$$\leq (Lip_e(\theta_{\xi_o}))^{N-1} \sigma_{\mathfrak{g}}\left(\theta_{\xi_o^{-1}}(T_{\xi_o}(\partial D)) \cap Box(r)\right),$$

$$(6.12) \qquad \sigma_{\mathfrak{g}}\left(\theta_{\xi_o}\left(\theta_{\xi_o^{-1}}(T_{\xi_o}(\partial D)) \cap Box(r)\right)\right)$$
$$\geq (Lip_e(\theta_{\xi_o^{-1}}))^{1-N} \sigma_{\mathfrak{g}}\left(\theta_{\xi_o^{-1}}(T_{\xi_o}(\partial D)) \cap Box(r)\right),$$

where we have denoted by $Lip_e(\theta_{\xi_o})$, $Lip_e(\theta_{\xi_o^{-1}})$ the Lipschitz constants, with respect to the Euclidean metric, of the maps θ_{ξ_o}, $\theta_{\xi_o^{-1}}$. It is clear that we can estimate such constants in terms of the Hilbert-Schmidt norms $||d\theta_{\xi_o}||$, $||d\theta_{\xi_o^{-1}}||$, where $d\theta_{\xi_o}$ is defined in (2.19). One the other hand, we have

$$||d\theta_{\xi_o}|| = \sqrt{m + k + \frac{1}{4}\sum_{j=1}^{m}\sum_{l=1}^{k} J(l,j)^2},$$

and therefore, there exists a constant $C^* = C^*(\mathfrak{g}, D) > 0$ (or, equivalently, $C^* = C^*(\boldsymbol{G}, \Omega)$, since $\Omega = \exp(D)$), such that

$$\sqrt{N} \leq ||d\theta_{\xi_o}|| \leq C^*, \qquad \text{for every} \quad \xi_o \in \partial D,$$

with an analogous estimate holding for $||d\theta_{\xi_o^{-1}}||$. Returning to (6.11), (6.12), we next want to estimate $\sigma_{\mathfrak{g}}\left(\theta_{\xi_o^{-1}}(T_{\xi_o}(\partial D)) \cap Box(r)\right)$. To this purpose, observe that

$\Pi = \theta_{\xi_o}^{-1}(T_{\xi_o}(\partial D))$ is a hyper-plane in $\mathfrak{g}$ passing through the origin. Our task is to identify the normal to Π at $0 = \theta_{\xi_o}^{-1}(\xi_o)$. Let $\nu(\xi_o)$ denote the normal to $T_{\xi_o}\partial D$ at ξ_o, that is, $\nu(\xi_o) = \nabla \rho(\xi_o)$, where $\rho = \phi \circ \exp$ is the defining function for ∂D. Let $\vec{N}$ denote the normal to Π at the origin, then by the chain rule we obtain

$$\vec{N} \;=\; \nabla(\rho \circ \theta_{\xi_o})(0) \;=\; d\theta_{\xi_o}(0)^t \left(\nabla \rho(\theta_{\xi_o}(0))\right) \;=\; d\theta_{\xi_o}(0)^t \left(\nabla \rho(\xi_o)\right) \;=\; d\theta_{\xi_o}(0)^t \left(\nu(\xi_o)\right).$$

From Proposition 2.2 we conclude

$$\vec{N} \;=\; d\theta_{\xi_o}(0)^t \left(\nu(\xi_o)\right) \;=\; \begin{pmatrix} Id_{m \times m} & J_{m \times k} \\ 0_{k \times m} & Id_{k \times k} \end{pmatrix} \nabla \rho(\xi_o)$$

$$= \begin{pmatrix} \frac{\partial \rho}{\partial x_1}(\xi_o) + \sum_{l=1}^{k} J(1,l) \frac{\partial \rho}{\partial y_l}(\xi_o) \\ \vdots \\ \frac{\partial \rho}{\partial x_m}(\xi_o) + \sum_{l=1}^{k} J(m,l) \frac{\partial \rho}{\partial y_l}(\xi_o) \\ \frac{\partial \rho}{\partial y_1}(\xi_o) \\ \vdots \\ \frac{\partial \rho}{\partial y_k}(\xi_o) \end{pmatrix} = \begin{pmatrix} X_1 \phi(g_o) \\ \vdots \\ X_m \phi(g_o) \\ Y_1 \phi(g_o) \\ \vdots \\ Y_k \phi(g_o) \end{pmatrix},$$

where in the last equality we have used Lemma 2.1. At this point we invoke the estimates in step one, which were obtained for a generic hyper-plane Π through the origin of $\mathfrak{g}$. Specifically, by choosing $(A, B) = (X\phi(g_o), Y\phi(g_o))$ in (6.6), and keeping (6.7) in mind, we finally reach the conclusion. This completes the proof. $\square$

6.3. Further analysis for Hörmander vector fields of step 2

We next establish some lower bounds for the function w which are crucial to the proof of Theorem 6.1. The following simple lemma is a key ingredient in the proof of Lemma 6.12 below.

LEMMA 6.7. *Let $X = \{X_1, ..., X_m\}$ be a system of C^∞ vector fields in $\mathbb{R}^n$, satisfying Hörmander's condition (1.4) at step 2. Let $\Omega \subset \mathbb{R}^n$ be a bounded, C^2 domain, with defining function ϕ. For every $x_o \in \Sigma = \Sigma_{\Omega, X}$ there exist indices $i_o, j_o \in \{1, ..., m\}$ such that*

(6.13) $$X_{i_o}(X_{j_o} \phi)(x_o) \;\neq\; 0.$$

Proof. Suppose (6.13) false, then there exists a $x_o \in \Sigma$ such that, for every $i, j \in \{1, ..., m\}$, one has

(6.14) $$X_i(X_j \phi)(x_o) \;=\; 0.$$

In this case,

$$<[X_i, X_j](x_o), \nabla \phi(x_o)> \;=\; [X_i, X_j]\phi(x_o) \;=\; X_i X_j \phi(x_o) - X_j X_i \phi(x_o) \;=\; 0.$$

This shows that $[X_i, X_j](x_o)$ is orthogonal to $\nabla \phi(x_o)$. Therefore,

$$\nabla \phi(x_o) \;\notin\; span\{X_i(x_o), [X_i, X_j](x_o), 1 \leq i, j \leq m\},$$

and hence, Hormander's condition is violated at x_o. We have reached a contradiction. $\square$

6.3. FURTHER ANALYSIS FOR HÖRMANDER VECTOR FIELDS OF STEP 2

To establish the next results, it will be convenient to make use of local coordinates and Taylor expansions. After a C^2 local change of variable, we can assume that the manifold $\partial \Omega = \mathcal{M}$ coincides with a portion of the hyper-plane $\mathcal{H} = \{(x_1, ..., x_n) \in \mathbb{R}^n \mid x_n = 0\}$. Thereby, a defining function for $\mathcal{M}$ is given by $\phi(x) = x_n$. We continue to indicate with the symbols $X_1, ..., X_m$ the transformed vector fields. Note that such vector fields are no longer C^∞ since the local diffeomorphism is only C^2. However, they still satisfy the finite rank condition (1.4) at step 2, since the latter is preserved by diffeomorphisms. In particular, the estimate (1.15) continues to hold with $\epsilon = 1/2$. As a consequence, Proposition 6.11 below remains valid for a system of C^2 vector fields satisfying the rank assumption at step 2, see [**Gro96**], pp.114-115. If we write the vector fields in the form

$$X_j = X_j(x) = \sum_{k=1}^{n} a_{j,k}(x) \partial_k, \qquad j = 1, ..., m,$$

then the angle function $w = |X\phi|$, defined in (3.3), is now given by

(6.15) $$w(x) = \left(\sum_{k=1}^{m} a_{k,n}(x_1, ..., x_{n-1}, 0)^2 \right)^{\frac{1}{2}},$$

whereas the characteristic set of $\mathcal{M}$ is

(6.16) $$\Sigma = \{x \in \mathcal{M} \mid a_{1,n}(x) = \cdots = a_{m,n}(x) = 0\}.$$

We let ∇' denote the $(n-1)$-dimensional gradient, taken with respect to the variables $x' = (x_1, ..., x_{n-1})$. We have the following.

LEMMA 6.8. *There exist $C_4 > 0$, a neighborhood $\Sigma_{\delta_1} = \{x \mid d(x, \Sigma) < \delta_1\}$ such that,*

$$\sum_{j=1}^{m} |\nabla' a_{j,n}(x)| \geq C_4 > 0$$

for all $x \in \Sigma_{\delta_1}$.

Proof. If $x_o \in \Sigma$, then noting that $X_{j_o} \phi = a_{j_o, n}$, the conclusion of Lemma 6.7 presently translates into the existence of i_o, j_o such that

$$0 \neq X_{i_o} a_{j_o, n}(x_o) = \sum_{k=1}^{n} a_{i_o, k}(x_o) \, \partial_k \, a_{j_o, n}(x_o)$$

$$= \sum_{k=1}^{n-1} a_{i_o, k}(x_o) \, \partial_k \, a_{j_o, n}(x_o) = <(\nabla' a_{j_o, n}(x_o), 0), X_{i_o}(x_o)>.$$

This shows that the vector $(\nabla' a_{j_o, n}(x_o), 0)$ has a non-zero component in span $\{X_1(x_o), ..., X_m(x_o)\}$. However, x_o is characteristic, and therefore span $\{X_1(x_o), ..., X_m(x_o)\} \subset T_{x_o} \mathcal{M} = \mathcal{H}$. As a consequence, we infer

(6.17) $$|\nabla' a_{j,n}(x_o)| \neq 0.$$

The conclusion now follows from (6.17), the continuity of $\sum_{j=1}^{m} |\nabla' a_{j,n}(x)|$, and the compactness of Σ. $\square$

Since the inequality that leads to the proof of Lemma 6.8 will be important later on, we state it in a separate lemma.

LEMMA 6.9. *There exists a neighborhood Σ_{δ_2} of Σ such that, for every $x_o \in \Sigma_{\delta_2}$, one can find indices $i_o, j_o \in \{1, ..., m\}$ for which*

(6.18) $$< (\nabla' a_{j_o,n}(x_o), 0), X_{i_o}(x_o) > \neq 0 .$$

PROOF. Since the manifold is bounded, Σ is compact. The lemma holds for each $y_o \in \Sigma$. By continuity, for each $y_o \in \Sigma$, there exists a neighborhood U_{y_o} such that $| < (\nabla' a_{j_o,n}(x), 0), X_{i_o}(x) > | > 0$ for every $x \in U_{y_o}$. By a standard covering argument, we obtain the desired neighborhood Σ_{δ_2}. □

The next proposition is an adaptation of [**Me93**, Lemma 2b], in which the author established an analogous result for the Heisenberg group $\mathbb{H}^n$.

PROPOSITION 6.10. *There exists $\beta > 0$, depending only on the manifold $\mathcal{M}$, and on the system $X = \{X_1, ..., X_m\}$, such that if $x_o \in \mathcal{M}$ is such that $w(x_o) > 0$, then*

$$\inf_{B(x_o, \beta\, w(x_o))} w > \frac{w(x_o)}{2} .$$

Proof. Recalling (6.15), and using a Taylor expansion for $a_{j,n}(x)$, we can write

$$w(x)^2 = \sum_{j=1}^{m} \Big(a_{j,n}(x_o) + \epsilon_j(|x - x_o|) \Big)^2$$

$$\geq \sum_{j=1}^{m} a_{j,n}(x_o)^2 + 2 \sum_{j=1}^{m} a_{j,n}(x_o)\, \epsilon_j(|x - x_o|),$$

where $\epsilon_j(|x - x_o|)$ denotes a function which tends to zero, as $x \to x_o$. It is then clear that there exists β', depending only on the system X, and on $\mathcal{M}$, such that if $|x - x_o| \leq (3/4)\beta' w(x_o)$, then

$$w(x)^2 \geq w(x_o)^2 - \beta'\, w(x_o)|x - x_o| \geq \frac{w(x_o)^2}{4} .$$

We have thus established the following

(6.19) $$\inf_{B_e(x_o, (3/4)\beta' w(x_o))} w > \frac{w(x_o)}{2},$$

where we have denoted by $B_e(x_o, R)$ the Euclidean ball centered at x_o with radius R. The passage from (6.19) to the statement in the theorem is now made possible by the crucial Proposition 1.10. The latter claims the existence of a bounded set $\tilde{U}$, containing $\mathcal{M}$, such that with $\tilde{C}$ given by (1.17), the inclusion (1.18) holds. It is thus clear that, by choosing

$$\beta = \frac{3\beta'}{4\tilde{C}} ,$$

the sought for conclusion follows. □

In the proof of the next lemma we will need the following result about non-holonomic geometry whose proof can be found in [**Gro96**], Sec.1.1.A', p.115.

PROPOSITION 6.11. *Let $c(t)$ be a smooth curve in $\mathbb{R}^n$ parametrized by arc-length, such that $c(0) = x_o$, and for which $c'(0) \in$ span $\{X_1(x_o), ..., X_m(x_o)\}$. There exists $T_o > 0$, and $\tilde{C}_1 > 0$, such that*

$$d(x_o, c(t)) \leq \tilde{C}_1\, t \qquad 0 \leq t \leq T_o.$$

6.3. FURTHER ANALYSIS FOR HÖRMANDER VECTOR FIELDS OF STEP 2

The following result plays an important role in the proof of Theorem 6.1.

LEMMA 6.12. *There exist C, $R_2 > 0$, depending on $\mathcal{M}$ and on X, such that for all $0 < r < R_2$, and every $x_o \in \mathcal{M}$, one has*

(6.20) $$\max_{B(x_o,r) \cap \mathcal{M}} w \geq C\, r.$$

Proof. If $\Sigma = \emptyset$, then the result is trivial, since the compactness of $\mathcal{M}$ gives
$$\inf_{\mathcal{M}} w \geq C,$$
for some $C > 0$ depending only on $\mathcal{M}$, and on X. To achieve (6.20), it thus suffices to choose $R_2 = C$. Consider next the case $\Sigma \neq \emptyset$. Let $\Sigma_\delta = U \cap \Sigma_{\delta_1} \cap \Sigma_{\delta_2}$ be a δ-neighborhood of Σ, where Σ_{δ_1}, Σ_{δ_2} are the neighborhoods from Lemmas 6.8 and 6.9 respectively and U is a neighborhood of Σ on which (1.15) holds. Since w is continuous, and only vanishes on Σ, we have
$$L = \inf_{\mathcal{M} \setminus \Sigma_\delta} w > 0.$$
Now, if we fix $R_2 < L$, then for $r < R_2$, and every $x_o \in \mathcal{M} \setminus \Sigma_\delta$, we have
$$\max_{B(x_o,r) \cap \mathcal{M}} w \geq w(x_o) \geq L > r.$$
We are left with considering the more subtle case $x_o \in \Sigma_\delta$. In what follows, to simplify the notation we set $a_j = a_{j,n}$. Using Taylor expansion around x_o, we obtain

(6.21)
$$w(x)^2 = \sum_{j=1}^{m} \Big[a_j(x_o) + \,<(\nabla' a_j(x_o), 0), x - x_o> +\, \epsilon_j(|x - x_o|) \Big]^2$$
$$= \sum_{j=1}^{m} \Big[a_j(x_o)^2 + \,<(\nabla' a_j(x_o), 0), x - x_o>^2 +\, \epsilon_j(|x - x_o|)^2$$
$$+ 2\, a_j(x_o) \,<(\nabla' a_j(x_o), 0), x - x_o> +\, 2\, a_j(x_o)\, \epsilon_j(|x - x_o|)$$
$$+ 2 \,<(\nabla' a_j(x_o), 0), x - x_o>\, \epsilon_j(|x - x_o|) \Big]$$
$$\geq \sum_{j=1}^{m} [a_j(x_o)^2 + \,<(\nabla' a_j(x_o), 0), x - x_o>^2 +\, \epsilon_j(|x - x_o|)^2$$
$$- 2\, |a_j(x_o)|\, |<(\nabla' a_j(x_o), 0), x - x_o>| -\, 2\, |a_j(x_o)|\, |\epsilon_j(|x - x_o|)|$$
$$- 2\, C\, |<(\nabla' a_j(x_o), 0), x - x_o>|\, |x - x_o|^2],$$

with $C = C(\mathcal{M}, X) > 0$. Suppose first that $x_o \in \Sigma$. (It is not necessary to distinguish the cases $x_o \in \Sigma$ and $x_o \in \Sigma_\delta \setminus \Sigma$ separately. However, for the clarity of the exposition, we prefer to make this distinction). Recalling (6.16), this assumption gives $a_j(x_o) = 0$, $j = 1, ..., n$. For every $x \in B(x_o, r) \cap \mathcal{M}$, (6.21) thus implies

(6.22)
$$w(x)^2 \geq \sum_{j=1}^{m} \Big\{ <(\nabla' a_j(x_o), 0), x - x_o>^2$$
$$- 2\, C\, |<(\nabla' a_j(x_o), 0), x - x_o>|\, |x - x_o|^2 +\, \epsilon_j(|x - x_o|)^2 \Big\}$$

$$\geq \sum_{j=1}^{m} [<(\nabla'a_j(x_o),0), x-x_o>^2 \ - \ 2\,C\,|<(\nabla'a_j(x_o),0), x-x_o>|\,|x-x_o|^2]$$

$$\geq \sum_{j=1}^{m} <(\nabla'a_j(x_o),0), x-x_o>^2 \ - \ C'\,r^3.$$

We notice that in obtaining the estimate $C'r^3$ for the term

$$2\,C\,|<(\nabla'a_j(x_o),0), x-x_o>|\,|x-x_o|^2$$

in the last inequality, we have used Proposition 1.10, which gives $B(x_o,r) \cap \mathcal{M} \subset B_e(x_o, \tilde{C}r) \cap \mathcal{M}$. Thereby, $|x-x_o| \leq \tilde{C}r$. To continue estimating from below, we invoke Lemma 6.9 to find indices i_o, j_o, for which $\nabla'a_{j_o}(x_o) \neq 0$, and (6.18) hold. Hence,

$$max \left\{ \sum_{j=1}^{m} <(\nabla'a_j(x_o),0), x-x_o>^2 \ \bigg|\ x \in B(x_o,r) \cap \mathcal{M} \right\}$$

$$\geq max \left\{ <(\nabla'a_{j_o}(x_o),0), x-x_o>^2 \ |\ x \in B(x_o,r) \cap \mathcal{M} \right\}.$$

Due to the nature of the constraint, it is not obvious at this point how to estimate from below the quantity in the right-hand side of the latter inequality. Our idea is to exploit Proposition 6.11 and make a clever choice of x which allows avoiding the actual computation of the constrained maximum. Without restriction, we can assume that $\tilde{C}_1 \geq 1$ in Proposition 6.11, whereas, by restricting the parameter R_2 further, we can suppose that $T_o \geq R_2$. Keeping (6.18) in mind, we fix $\lambda > 0$ so that

$$\lambda\,\tilde{C}_1 \frac{|<X_{i_o}(x_o), (\nabla'a_{j_o}(x_o),0)> X_{i_o}(x_o)|}{|\nabla'a_{j_o}(x_o)|} \ = \ 1,$$

with $\tilde{C}_1$ being the constant in Proposition 6.11. We now define a point $\bar{x}$ by the equation

$$(6.23) \qquad \bar{x} \ - \ x_o \ = \ \frac{\lambda r}{|\nabla'a_{j_o}(x_o)|} \ <X_{i_o}(x_o), (\nabla'a_{j_o}(x_o),0)> X_{i_o}(x_o).$$

Let us observe that, by our choice of λ, and of R_2, we have

$$(6.24) \qquad |\bar{x} - x_o| \ = \ \frac{r}{\tilde{C}_1} \ \leq \ r \ < \ R_2 \ \leq \ T_o\,.$$

We next define the smooth curve

$$(6.25) \qquad c(t) \ = \ x_o \ + \ t\,\frac{\bar{x} - x_o}{|\bar{x} - x_o|}, \qquad 0 \leq t \leq 1.$$

Clearly, $c(t)$ is parametrized by arc-length, and $c(0) = x_o$. By Proposition 6.11, we have

$$d(x_o, \bar{x}) \ \leq \ \tilde{C}_1\,t_o,$$

where t_o is such that $c(t_o) = \bar{x}$. Of course, we need to make sure that $t_o = |\bar{x} - x_o| \leq T_o$, but this is guaranteed by (6.24). We conclude

$$d(x_o, \bar{x}) \ \leq \ C\,|\bar{x} - x_o| \ \leq \ r.$$

This proves that $\bar{x} \in B(x_o, r)$. On the other hand, from (6.23), and from $x_o \in \Sigma$ (see (6.16)), it is clear that $\bar{x} \in \mathcal{M}$, therefore we have

$$\bar{x} \ \in B(x_o, r) \ \cap \ \mathcal{M}.$$

6.3. FURTHER ANALYSIS FOR HÖRMANDER VECTOR FIELDS OF STEP 2

This important conclusion allows to continue as follows

(6.26)
$$max\ \{<(\nabla'a_{j_o}(x_o),0), x-x_o>^2\ |\ x \in B(x_o,r) \cap \mathcal{M}\}$$
$$\geq\ <(\nabla'a_{j_o}(x_o),0), \bar{x}-x_o>^2\ =\ \left(\lambda\ \frac{<X_{i_o}(x_o),(\nabla'a_{j_o}(x_o),0)>^2}{|\nabla'a_{j_o,n}(x_o)|}\right)^2 r^2\ =\ C^*\ r^2,$$

where in the second to the last equality we have used (6.23). Exploiting (6.17), and (6.18), we conclude that $C^* = C^*(\mathcal{M},X) > 0$. At this point we insert (6.26) in (6.22). By choosing $R_2 > 0$ sufficiently small, depending on the constants C^*, C' in these estimates, we conclude the existence of $C = C(\mathcal{M},X) > 0$ such that, for every $x_o \in \Sigma$, and for any $0 \leq r \leq R_2$, one has

$$max\{w(x)\ |\ x \in B(x_o,r) \cap \mathcal{M}\}\ \geq\ C\,r\,.$$

This proves the lemma when x_o is characteristic. Finally, we consider the case $x_o \in \Sigma_\delta \setminus \Sigma$. Let $\tilde{C}_2$ be the constant in (1.15) (with $\epsilon = 1/2$ in (1.15)). Let i_o, j_o be given by Lemma 6.9 and let $\tilde{\lambda}$ be such that

$$\tilde{\lambda}\,\tilde{C}_1\,\frac{|<X_{i_o}(x_o),(\nabla'a_{j_o}(x_o),0)>X_{i_o}(x_o)|}{|\nabla'a_{j_o}(x_o)|}\ =\ \frac{1}{2}.$$

Next, we would like to choose a candidate $\bar{u} \in B(x_o,r) \cap \mathcal{M}$ such that $w(\bar{u}) \geq C^*r$. When $x_o \in \Sigma_\delta \setminus \Sigma$, $\bar{u}$ has to be choosen slightly differently and carefully. As before, let $\bar{x}$ be defined by (6.23) with $\tilde{\lambda}$ playing the role of λ there. By considering the curve defined in (6.25), using our new choice of $\tilde{\lambda}$ and arguing as before we now have

(6.27)
$$d(x_o,\bar{x})\ \leq\ \tilde{C}_1 t_o\ =\ \tilde{C}_1\,|\bar{x}-x_o|\ =\ \frac{r}{2}.$$

We are ready to specify our choice of $\bar{u}$.

$$\bar{u}\ =\ x_o\ +\ \frac{\tilde{\lambda}r}{|\nabla'a_{j_o}(x_o)|}\ <X_{i_o}(x_o),(\nabla'a_{j_o}(x_o),0)>(a_{i_o,1}(x_o),...,a_{i_o,n-1},0).$$

That is, $\bar{u}$ is the orthogonal projection of our new choice of $\bar{x}$ onto $\mathcal{M}$. Our next task is to show that $\bar{u} \in B(x_o,r)$ (hence $\bar{u} \in B(x_o,r) \cap \mathcal{M}$). Clearly, as before, by exploiting (6.17) and (6.18) we obtain $\bar{C} > 0$ such that

(6.28)
$$|<(\nabla'a_{j_o}(x_o),0),\bar{x}-x_o>|\ \geq\ \bar{C}\,r.$$

To continue, let

$$A\ =\ \frac{1}{4}\,min(\bar{C},\frac{\tilde{C}_1}{\tilde{C}_2^2}\,inf\{|X_{j_o}(x)|\,|\,x \in \overline{\Sigma_\delta},\})\ >\ 0$$

where in the above, the indices j_o are the ones obtained from Lemma 6.9. We assume $w(x_o) < A\,r$. Otherwise, the conclusion of the Lemma follows trivially. With this assumption in hand, using our choices of A, $\tilde{\lambda}$ and recalling (6.15) we obtain

$$
\begin{aligned}
(6.29) \quad \tilde{C}_2 |\bar{x} - \bar{u}|^{\frac{1}{2}} &= \tilde{C}_2 \left(\frac{\tilde{\lambda} r}{|\nabla' a_{j_o}(x_o)|} | < (\nabla' a_{j_o}(x_o), 0), X_{i_o}(x_o) > ||a_{i_o}(x_o)| \right)^{\frac{1}{2}} \\
&= \tilde{C}_2 \left(\frac{r}{2\tilde{C}_1 |X_{i_o}(x_o)|} |a_{i_o}(x_o)| \right)^{\frac{1}{2}} \\
&\leq \tilde{C}_2 \left(\frac{A r^2}{2\tilde{C}_1 |X_{i_o}(x_o)|} \right)^{\frac{1}{2}} < \frac{r}{2}.
\end{aligned}
$$

Hence

$$
\begin{aligned}
d(x_o, \bar{u}) &\leq d(x_o, \bar{x}) + d(\bar{x}, \bar{u}) \\
\text{(by (1.15))} &\leq d(x_o, \bar{x}) + \tilde{C}_2 |\bar{x} - \bar{u}|^{\frac{1}{2}} \\
\text{(by (6.27) and (6.29))} &\leq \frac{r}{2} + \frac{r}{2} = r.
\end{aligned}
$$

This shows $\bar{u} \in B(x_o, r)$. To complete the proof, our goal is to show $w(\bar{u}) \geq Cr$ for some $C = C(X, \mathcal{M}) > 0$. To this end observe that since the last component of $(\nabla' a_{j_o}(x_0), 0)$ is zero we obtain

(6.30)
$$
\begin{aligned}
&< (\nabla' a_{j_o}(x_0), 0), \bar{u} - x_o >^2 - 2|a_{j_o}(x_o)| \, | < (\nabla' a_{j_o}(x_0), 0), \bar{u} - x_o > | \\
&= < (\nabla' a_{j_o}(x_0), 0), \bar{x} - x_o >^2 - 2|a_{j_o}(x_o)| \, | < (\nabla' a_{j_o}(x_0), 0), \bar{x} - x_o > | \\
\text{(by (6.28))} &\geq \bar{C}(\bar{C} - 2A)r^2 \geq \frac{\bar{C}^2}{2} r^2
\end{aligned}
$$

by our choice of A. Using the estimate of (6.30) in (6.21) and arguing as in the case where $x_o \in \Sigma$ we obtain $w(x_o) > Cr$ and reach the conclusion. $\square$

6.4. Second proof of Theorem 6.1

Given a Carnot group G of step 2, let $\Omega \subset G$ be a bounded C^2 domain with characteristic set $\Sigma = \Sigma_{\Omega, X}$. We set $R'_1 = min(R_1, R_2)$, where R_1 and R_2 are the parameters in Lemmas 6.6 and 6.12 respectively. With $g_o \in \partial\Omega$, and $0 < r < R'_1$, we consider the surface ball $B(g_o, r) \cap \partial\Omega$. Let $g_1 \in B(g_o, r/2) \cap \partial\Omega$ be such that

$$(6.31) \qquad w(g_1) = max\{w(g) \, | \, g \in B(g_o, r/2) \cap \partial\Omega\}.$$

We now claim that $w(g_1) > 0$. In fact, if by contradiction $w(g_1) = 0$, then we must have

$$(6.32) \qquad B(g_o, r/2) \cap \partial\Omega \subset \Sigma.$$

Thanks to (1.13), and to Theorem 1.7, which in the present case of a group of step 2 holds with $\epsilon = 1/2$, we conclude the existence of a constant $C > 0$, depending on G and Ω, such that for every $g, g' \in \overline{\Omega}$

$$d(g, g') \leq C \, d_e(g, g')^{1/2},$$

6.4. SECOND PROOF OF THEOREM 6.1

where we have denoted by $d_e(g, g')$ the Riemannian distance in $\boldsymbol{G}$. This inequality, combined with (6.32), shows that
$$B_e(g_o, (r/C)^2) \cap \partial\Omega \subset \Sigma.$$
This, however, contradicts Theorem 3.2, and therefore the claim is proved. We can thus invoke Proposition 6.10, obtaining the existence of $\beta > 0$ such that, for all $g \in B(g_1, \beta\, w(g_1)) \cap \partial\Omega$, we have

$$(6.33) \qquad w(g) \geq \frac{w(g_1)}{2} > 0.$$

We now distinguish two cases.

Case 1: $w(g_1) \leq \frac{3}{2\beta} r$ (the "nearly characteristic" case). In this situation, the hypothesis $g_1 \in B(g_o, r/2) \cap \partial\Omega$ implies
$$B(g_1, \frac{\beta}{3} w(g_1)) \cap \partial\Omega \subset B(g_o, r) \cap \partial\Omega.$$

This gives,

$$(6.34) \qquad \mu(B(g_o, r) \cap \partial\Omega) = \int_{B(g_o, r) \cap \partial\Omega} w(g)\, d\sigma(g)$$

$$\geq \int_{B(g_1, \frac{\beta}{3} w(g_1)) \cap \partial\Omega} w(g)\, d\sigma(g)$$

$$\text{(by (6.33))} \quad \geq \frac{w(g_1)}{2} \int_{B(g_1, \frac{\beta}{3} w(g_1)) \cap \partial\Omega} d\sigma(g)$$

$$\text{(by Lemma 6.6)} \quad \geq C_1 \frac{w(g_1)}{2} \frac{(\frac{\beta}{3} w(g_1))^{Q-1}}{\eta(g_1, \frac{\beta}{3} w(g_1))}.$$

From the definition (6.2) of the function η we see that, if
$$\eta\left(g_1, \frac{\beta}{3} w(g_1)\right) = \frac{w(g_1)}{|\nabla\phi(g_1)|},$$
then (6.34) reduces to
$$\mu(B(g_o, r) \cap \partial\Omega) \geq C\, w(g_1)^{Q-1} \geq C\, r^{Q-1},$$
where in the last inequality we have used Lemma 6.12. The crucial role of the latter in this last step cannot be emphasized enough. If, instead,
$$\eta\left(g_1, \frac{\beta}{3} w(g_1)\right) = \frac{\beta}{3} w(g_1) \frac{|Y\phi(g_1)|}{|\nabla\phi(g_1)|},$$
then, in particular, we have

$$(6.35) \qquad \eta\left(g_1, \frac{\beta}{3} w(g_1)\right) \leq \frac{\beta}{3} w(g_1).$$

This follows from the inequality $|Y\phi| \leq |\nabla\phi|$, which can be easily proved observing that, in the exponential coordinates of $\boldsymbol{G}$, one has $Y_s = \partial/\partial y_s$. Inserting (6.35) in (6.34), we obtain
$$\mu(B(g_o, r) \cap \partial\Omega) \geq C\, w(g_1)^{Q-1} \geq C\, r^{Q-1}.$$
where, again, we have used Lemma 6.12.

Case 2: $w(g_1) > \frac{3}{2\beta} r$ (the "non-characteristic" case). In this situation, we have
$$B(g_o, r) \cap \partial\Omega \subset B(g_1, \beta w(g_1)) \cap \partial\Omega,$$

and therefore (6.33) implies $w(g) \geq w(g_1)/2$ for all $g \in B(g_o, r) \cap \partial\Omega$. This gives

$$(6.36) \qquad \mu(B(g_o, r) \cap \partial\Omega) \geq \frac{w(g_1)}{2} \sigma(B(g_o, r) \cap \partial\Omega) \geq C \frac{w(g_1)}{2\eta(g_o, r)} r^{Q-1},$$

by Lemma 6.6 again. If $\eta(g_o, r) = w(g_o)/|\nabla\phi(g_o)|$, then (6.31) implies that

$$\frac{w(g_1)}{\eta(g_o, r)} \geq C'' > 0,$$

thus the conclusion of the theorem follows. If, instead, $\eta(g_o, r) = r\,|Y\phi(g_o)|/|\nabla\phi(g_o)|$, then as for (6.35) we obtain

$$\mu(B(g_o, r) \cap \partial\Omega) \geq \frac{C}{2r} w(g_1) \frac{|\nabla\phi(g_o)|}{|Y\phi(g_o)|} r^{Q-1} > \frac{3C}{4\beta} r^{Q-1},$$

where in the last inequality we have used the assumption that $w(g_1) > \frac{3}{2\beta}r$. This completes the proof.

REMARK 6.13. In connection with Theorem 6.2 we mention a remarkable result by Ambrosio [**Am01**]. The latter states, among other things, that the generalized perimeter $P(\cdot;\cdot)$ in a complete metric space (S, d) is *asymptotically doubling*. That is, if $\Omega \subset S$ is any set such that $P(\Omega; S) < \infty$, then

$$(6.37) \qquad \limsup_{t \to 0} \frac{P(\Omega, B(x, 2t))}{P(\Omega, B(x, t))} < \infty \qquad \text{for} \quad P(\Omega; \cdot) - \text{a.e.}\, x \in S.$$

The asymptotic doubling (6.37) suffices to obtain a Vitali type covering theorem with respect to the measure $P(E;\cdot)$, as in [**Fe69**], Theorem 2.8.17, and thereby develop a theory of differentiation a la De Giorgi.

The basic assumptions in [**Am01**] are that: a) (S, d) be a k-Ahlfors space, i.e., that S be endowed with a Borel measure ν such that for some $a > 0$ one has for every $x \in S$, and any $r > 0$

$$(6.38) \qquad a\, r^k \leq \nu(B(x, r)) \leq a^{-1}\, r^k.$$

Secondly, (S, d) supports a Poincaré inequality, with a gradient according to Hajlasz, or to Heinonen and Koskela, see [**HK98**]. With these assumptions in force, one can introduce a notion of perimeter and develop a theory of first-order Sobolev spaces similarly to what was done in [**GN96**] for CC spaces, see [**Mi00**]. If **G** is a Carnot group with homogeneous dimension Q, then (6.38) is valid with $d\nu = dg$ and $k = Q$, see (2.17). As a consequence, (6.37) holds in any Carnot group. The assumption (6.38) leaves out the case of vector fields of Hörmander type. However, recently the same author [**Am02**] has established (6.37) under more general assumptions which include this setting. As we have seen in this chapter the passage from the asymptotic information in (6.37), to the precise Ahlfors type estimates in Theorem 6.2 hides a considerable amount of work. This is unfortunately unavoidable since, in the applications of CC geometry to partial differential equations, one needs to confront the essential obstacle due to the presence of characteristic points on the boundary of a given domain Ω. In this perspective, it becomes necessary to develop a more precise quantitative analysis of the interplay between the questions at study, and the geometry of the ambient space.

6.5. Failure of the 1-Ahlfors condition for the X-perimeter of $C^{1,\alpha}$ domains

In Theorem 6.1 we have assumed that the relevant domain be of class $C^{1,1}$. We prove here that it is not possible to further weaken the regularity hypothesis on the domain. In other words, the $C^{1,1}$ smoothness of the domain Ω in Theorems 6.1 and 6.2 is best possible for the X-perimeter measure $P_X(\Omega; \cdot)$ to satisfy the 1-Ahlfors condition. Consider in fact the first Heisenberg group $\mathbb{H}^1$, denote with $g = (x, y, t)$ the generic point, and let $z = (x, y)$. We indicate with $e = (0, 0, 0)$ the group identity. Recall that the homogeneous dimension of $\mathbb{H}^1$ associated with the parabolic dilations $\delta_\lambda(g) = (\lambda z, \lambda^2 t)$ is $Q = 4$. Given any $\alpha \in (0, 1)$, with $\beta = 1 + \alpha$, we consider the bounded domain

(6.39) $$\Omega = \{g \in \mathbb{H}^1 \mid 0 < t < 1,\ t > |z|^\beta\}.$$

Except for the intersection with the plane $\{g \in \mathbb{H}^1 \mid t = 1\}$ (which we could have as well rounded off without introducing new characteristic points), the boundary of Ω is a manifold of class $C^{1,\alpha}$. Set $F = \partial\Omega \setminus \{g \in \partial\Omega \mid t = 1\}$, then e is the only characteristic point of F. We will prove that

(6.40) $$\lim_{r \to 0} \frac{P_X(\Omega; B(e, r))}{r^3} = 0.$$

Keeping in mind that $|B(g, r)|/r \cong r^3$, we conclude from (6.40) that for no $M > 0$, and $0 < R_o < 1$, can the inequality

$$P_X(\Omega; B(e, r)) \geq M^{-1} \frac{|B(e, r)|}{r},\qquad 0 < r < R_o,$$

possibly hold. Therefore, $P_X(\Omega; \cdot)$ is not a lower 1-Ahlfors measure (although it can be proved that it is an upper 1-Ahlfors measure). To establish (6.40) it suffices to prove that

(6.41) $$\lim_{r \to 0} \frac{\int_{F \cap Box(e,r)} |X\phi(g)|\,d\sigma(g)}{r^3} = 0,$$

where $Box(e, r) = \{g \in \mathbb{H}^1 \mid |z| < r, |t| < r^2\}$, and $\phi(g) = |z|^\beta - t$ is a defining function for Ω near e. In fact, thanks to (4.8) we have for $0 < r < 1$

$$P_X(\Omega; B(e, r)) = \int_{F \cap B(e,r)} |X\phi(g)|\,d\sigma(g),$$

and Theorem 1.17 gives

$$\int_{F \cap Box(e,\delta^{-1}r)} |X\phi(g)|\,d\sigma(g) \leq \int_{F \cap B(e,r)} |X\phi(g)|\,d\sigma(g) \leq \int_{F \cap Box(e,\delta^{-1}r)} |X\phi(g)|\,d\sigma(g)$$

for some $\delta > 0$ independent of r. From (3.19) se find

$$X_1\phi(g) = \beta\, x\, |z|^{\beta-2} - 2\,y, \qquad X_2\phi(g) = \beta\, y\, |z|^{\beta-2} + 2\,x.$$

Since $1 < \beta < 2$, it is easy to recognize that for every $g \in F$ we have

(6.42) $$\beta\, |z|^{\beta-1} \leq |X\phi(g)| \leq \sqrt{4 + \beta^2}\, |z|^{\beta-1}.$$

Noting that $d\sigma = \sqrt{1 + \beta^2 |z|^{2(\beta-1)}}\,dz$, we have

$$dz \leq d\sigma(g) \leq \sqrt{1 + \beta^2}\,dz,\qquad\text{on}\ \ F.$$

Again the condition $1 < \beta < 2$, and (6.42), imply that

$$(6.43) \qquad \int_{F \cap Box(e,r)} |X\phi(g)| \, d\sigma(g) \cong \int_{|z|<r^{2/\beta}} |z|^{\beta-1} \, dz \cong r^{2+2/\beta} \ .$$

The latter equation immediately implies (6.41). What goes on with this example is that, despite the $C^{1,\alpha}$ smoothness of the boundary, the parabolic dilations of the group actually turn the domain Ω into a "cuspidal" region, when seen with the sub-Riemannian glasses. We will return to this example in Proposition 10.8 in Chapter 10.

CHAPTER 7

Fine differentiability properties of Sobolev functions

As we have seen in Chapters 5 and 6, in the study of the traces the measure μ in Definition 1.3 is supported on a lower dimensional manifold, and therefore on a set of Lebesgue measure zero. Thereby, a function in L^p with respect to Lebesgue measure may not see the support of μ. This is a serious obstacle in the study of trace inequalities since, a priori, one only knows membership of the appropriate functions in a sub-elliptic Sobolev space with respect to Lebesgue measure. This chapter is devoted to by-passing this obstacle. Our main result is Theorem 7.8, which constitutes a refinement of Lebesgue differentiation theorem.

7.1. Poincaré inequality, fractional integrals and improved representation formulas

In the theory of partial differential equations a classical inequality is that of Poincaré. Combined with the doubling condition (1.22), such inequality represents nowadays a basic constituent of analysis and geometry. This aspect was, independently, first emphasized by the works of Grigor'yan [**Gri91**] and Saloff-Coste [**SaCo92**]. Subsequently, the works of several people have further developed these ideas, see, e.g., [**HK98**], [**G02**], and the references therein. In the setting of CC metrics generated by a system of Hörmander type X, an inequality of Poincaré type was proved by D. Jerison in his fundamental paper [**J86**]. An earlier version for Carnot groups was obtained by Varopoulos [**Va86**]. More recently, an interesting different approach has been proposed by Lanconelli and Morbidelli in [**LM00**]. Henceforth, for a given measure $d\nu$ on $\mathbb{R}^n$, $1 \leq p \leq \infty$, and an open set $\Omega \subset \mathbb{R}^n$, we define

(7.1) $\quad \mathcal{L}^{1,p}(\Omega, d\nu) = \{f \in L^p(\Omega, d\nu) \mid X_j f \in L^p(\Omega, d\nu), \ j = 1, ..., m\}.$

Such space is endowed with the obvious norm

$$||f||_{\mathcal{L}^{1,p}(\Omega,d\nu)} = ||f||_{L^p(\Omega,d\nu)} + ||Xf||_{L^p(\Omega,d\nu)}.$$

Here is Jerison's Poincaré inequality [**J86**].

THEOREM 7.1. *Let $U \subset \mathbb{R}^n$ be a bounded set. There exist constants $C_2, R_o > 0$, depending on U and on X, such that for any $x_o \in U$, $0 < r \leq R_o$, and $f \in \mathcal{L}^{1,1}(B(x_o, r), dx)$, one has*

$$\int_{B(x_o,r)} |f(x) - f_{B(x_o,r)}| \, dx \leq C_2 \, r \int_{B(x_o,r)} |Xf(x)| \, dx.$$

A basic consequence of Theorem 7.1 is the following representation formula established independently in [**FLW95**], [**CDG97**], see also [**FLW96**].

THEOREM 7.2. *Given a bounded set $U \subset \mathbb{R}^n$, there exist constants $\delta \geq 1, C > 0$, depending on U and on X, such that for any $x_o \in U$, $0 < r \leq R_o$, and $f \in \mathcal{L}^{1,1}(B(x_o, \delta r), dx)$, one has*

$$(7.2) \qquad |f(x) - f_{B(x_o,r)}| \leq C \int_{B(x_o, \delta r)} |Xf(y)| \frac{d(x,y)}{|B(x, d(x,y))|} \, dy \,,$$

whenever $x \in B(x_o, r)$ is such that

$$\lim_{t \to 0} \frac{1}{|B(x,t)|} \int_{B(x,t)} f(y) \, dy = f(x).$$

In particular, (7.2) holds for dx-a.e. $x \in B(x_o, r)$.

For the rest of the chapter, δ will have the same meaning as in Theorem 7.2. One aspect of Theorem 7.2 which represents a problem in the study of traces is the fact that the exceptional set is with respect to the Lebesgue measure dx. Unfortunately, we do not know how to relate sets of dx-measure zero to negligible sets with respect to a different measure. For this reason, we need to refine the statement of Theorem 7.2. By exploiting the assumption that f belong to a Sobolev space, we remove the dx-a.e. part of its conclusion, and also introduce an improving factor. But first, we introduce two definitions.

DEFINITION 7.3. *Let $U \subset \mathbb{R}^n$ be a bounded set, with characteristic local parameters C_1, R_o, and with local homogeneous dimension $Q = \log_2 C_1$. For every $0 < \alpha \leq Q$, and every $0 < R \leq R_o$, the operator of fractional integration of order α, relative to the ball $B(x_o, \delta R)$, where $x_o \in U$, is defined by*

$$I_\alpha \, f(x) = \int_{B(x_o, \delta R)} |f(y)| \frac{d(x,y)^\alpha}{|B(x, d(x,y))|} \, dy, \qquad x \in B(x_o, \delta R).$$

Using this definition, we can reformulate (7.2) as follows

$$(7.3) \qquad |f(x) - f_{B(x_o,r)}| \leq C \, I_1(|Xf|)(x) \qquad \text{for} \quad dx - \text{a.e. } x \in B(x_o, r).$$

DEFINITION 7.4. *For a function $f \in L^1_{loc}(\mathbb{R}^n, dx)$ we let*

$$f^\star(x) = \begin{cases} \lim_{t \to 0} \frac{1}{|B(x,t)|} \int_{B(x,t)} f(y) \, dy & \text{if this limit exists,} \\ 0 & \text{otherwise.} \end{cases}$$

We call $f^\star$ the *precise representation* of f.

It is easy to verify that the function $f^\star$ has the following elementary properties.
1. If $f = g$ dx-a.e., then $f^\star = g^\star$ everywhere.
2. Lebesgue differentiation theorem for spaces of homogeneous type [**C76**] implies $f = f^\star$ dx-a.e.
3. If for some $p \geq 1$ we have $f \in \mathcal{L}^{1,p}(B(x_o, R), dx)$, then $X_j f^\star = (X_j f)^\star$, and moreover $f^\star \in \mathcal{L}^{1,p}(B(x_o, R), dx)$ also.

7.1. POINCARÉ INEQUALITY, FRACTIONAL INTEGRALS AND IMPROVED...

THEOREM 7.5. *Let $U \subset \mathbb{R}^n$ be bounded, with local parameters C_1, R_o. For any $0 < \alpha < 1$, there exists $C = C(C_1, \alpha) > 0$ such that, for any $x_o \in U$, $0 < R \leq R_o$, and $f \in \mathcal{L}^{1,1}(B(x_o, \delta R), dx)$, one has*

$$|f^\star(x) - f^\star_{B(x,r)}| \leq C\, r^{1-\alpha}\, I_\alpha(|Xf^\star|)(x)$$

for every $x \in B(x_o, R)$, and $0 < r < \mathrm{dist}(x, \partial B(x_o, \delta R))$.

Proof. We fix $x \in B(x_o, R)$, and $0 < r < \mathrm{dist}(x, \partial B(x_o, \delta R))$. Our goal is to prove that, for $0 < \eta < \sigma \leq r$, the following estimate holds
(7.4)
$$|f_{B(x,\eta)} - f_{B(x,\sigma)}| \leq C\, \eta^{1-\alpha}\, I_\alpha(|Xf|)(x) + C\,(\sigma^{1-\alpha} - 2^{-(1-\alpha)}\, \eta^{1-\alpha})\, I_\alpha(|Xf|)(x),$$

for some constant C, depending only on C_1 and α. Suppose for a moment that we have proved (7.4). We distinguish two cases. If $I_\alpha(|Xf^\star|)(x) = I_\alpha(|Xf|)(x) = \infty$, then the inequality in the theorem is trivially true. If, instead, $I_\alpha(|Xf|)(x) < \infty$, we infer from (7.4) that $\{f_{B(x,t)}\}_{0<t<r}$ is a Cauchy sequence, hence

$$\lim_{t \to 0} \frac{1}{|B(x,t)|} \int_{B(x,t)} f(y)\, dy$$

exists. By Definition 7.4, one has

$$f^\star(x) = \lim_{t \to 0} \frac{1}{|B(x,t)|} \int_{B(x,t)} f(y)\, dy.$$

Noting that $f_{B(x,r)} = f^\star_{B(x,r)}$, passing to the limit as $\eta \to 0$ in (7.4), and choosing $\sigma = r$, we obtain the conclusion. To complete the proof of the theorem, we are thus left with proving (7.4). If $N \in \mathbb{N}$, we have

$$|f_{B(x,2^{-N-1}\sigma)} - f_{B(x,\sigma)}| \leq \sum_{k=0}^{N} |f_{B(x,2^{-k-1}\sigma)} - f_{B(x,2^{-k}\sigma)}|$$

$$\leq \sum_{k=0}^{N} \frac{1}{|B(x, 2^{-k-1}\sigma)|} \int_{B(x, 2^{-k-1}\sigma)} |f(y) - f_{B(x, 2^{-k}\sigma)}|\, dy$$

(by Theorem 1.12) $\leq C \sum_{k=0}^{N} \frac{1}{|B(x, 2^{-k}\sigma)|} \int_{B(x, 2^{-k}\sigma)} |f(y) - f_{B(x, 2^{-k}\sigma)}|\, dy$

(by Theorem 7.1) $\leq C \sum_{k=0}^{N} \frac{2^{-k}\sigma}{|B(x, 2^{-k}\sigma)|} \int_{B(x, \delta 2^{-k}\sigma)} |Xf(y)|\, dy$

$$= C \sum_{k=0}^{N} (2^{-k}\sigma)^{1-\alpha} \frac{(2^{-k}\sigma)^\alpha}{|B(x, 2^{-k}\sigma)|} \int_{B(x, \delta 2^{-k}\sigma)} |Xf(y)|\, dy$$

(by Proposition 1.15) $\leq C\, \delta^{-\alpha}\, \sigma^{1-\alpha}$

$$\sum_{k=0}^{N} 2^{-k(1-\alpha)} \int_{B(x_o, \delta R)} |Xf(y)|\, \frac{d(x,y)^\alpha}{|B(x, \frac{d(x,y)}{\delta})|}\, dy$$

(by Theorem 1.12) $\leq C\, \sigma^{1-\alpha}\, I_\alpha(|Xf|)(x) \sum_{k=0}^{N} 2^{-k(1-\alpha)}$

$$\leq C\, \sigma^{1-\alpha}\, \frac{1 - 2^{-(1-\alpha)(N+1)}}{1 - 2^{-(1-\alpha)}}\, I_\alpha(|Xf|)(x).$$

We next choose $N \in \mathbb{N}$ such that $2^{-N-1}\sigma \leq \eta < 2^{-N}\sigma$. Then
$$1 - 2^{-(1-\alpha)(N+1)} \leq 1 - 2^{-(1-\alpha)} \left(\frac{\eta}{\sigma}\right)^{1-\alpha}.$$
Using this fact in the right-hand side of the above string of inequalities, we obtain
$$(7.5) \quad |f_{B(x,2^{-N-1}\sigma)} - f_{B(x,\sigma)}| \leq C \left(\sigma^{1-\alpha} - 2^{-(1-\alpha)}\eta^{1-\alpha}\right) I_\alpha(|Xf|)(x).$$
On the other hand, we find

(7.6)
$$|f_{B(x,\eta)} - f_{B(x,2^{-N-1}\sigma)}| \leq \frac{1}{|B(x,2^{-N-1}\sigma)|} \int_{B(x,2^{-N-1}\sigma)} |f(y) - f_{B(x,\eta)}| \, dy$$
(by Theorem 1.12 and the choice of N)
$$\leq C \frac{1}{|B(x,\eta)|} \int_{B(x,\eta)} |f(y) - f_{B(x,\eta)}| \, dy$$
(by Theorem 7.1)
$$\leq C \frac{\eta}{|B(x,\eta)|} \int_{B(x,\delta\eta)} |Xf(y)| \, dy$$
(by Proposition 1.15 and Theorem 1.12)
$$\leq C \, \eta^{1-\alpha} I_\alpha(|Xf|)(x).$$

Using (7.5), and (7.6), we conclude
$$|f_{B(x,\eta)} - f_{B(x,\sigma)}|$$
$$\leq |f_{B(x,\eta)} - f_{B(x,2^{-N-1}\sigma)}| + |f_{B(x,2^{-N-1}\sigma)} - f_{B(x,\sigma)}|$$
$$\leq C \, \eta^{1-\alpha} I_\alpha(|Xf|)(x) + C \left(\sigma^{1-\alpha} - 2^{-(1-\alpha)}\eta^{1-\alpha}\right) I_\alpha(|Xf|)(x).$$
This proves (7.4), and completes the proof. $\square$

An immediate consequence of Theorem 7.5 is the following.

THEOREM 7.6. *Let U, C_1, R_o, α, be as in Theorem 7.5. There exists $C = C(C_1, \alpha) > 0$ such that for every $x_o \in U$, $0 < R \leq R_o$, and $f \in \mathcal{L}^{1,1}(B(x_o, \delta R)$, one has*
$$\frac{1}{|B(x,r)|} \int_{B(x,r)} |f^\star(x) - f^\star(y)| \, dy \leq C \, r^{1-\alpha} I_\alpha(|Xf^\star|)(x)$$
for every $x \in B(x_o, R)$, and $0 < r < dist(x, \partial B(x_o, \delta R))$.

PROOF. Theorems 7.2, 7.5 imply
$$\frac{1}{|B(x,r)|} \int_{B(x,r)} |f^\star(x) - f^\star(y)| \, dy$$
$$\leq |f^\star(x) - f^\star_{B(x,r)}| + \frac{1}{|B(x,r)|} \int_{B(x,r)} |f^\star(y) - f^\star_{B(x,r)}| \, dy$$
$$\leq C \, r^{1-\alpha} I_\alpha(|Xf^\star|)(x) + C \frac{1}{|B(x,r)|} \int_{B(x,r)} \left(\int_{B(x,\delta r)} |Xf^\star(z)| \frac{d(y,z)}{|B(y,d(y,z))|} \, dz\right) dy$$

The second term in the right-hand side of the latter inequality is estimated as follows

$$\frac{1}{|B(x,r)|}\int_{B(x,r)}\left(\int_{B(x,\delta r)}|Xf^\star(z)|\frac{d(y,z)}{|B(y,d(y,z))|}dz\right)dy$$

$$= \frac{1}{|B(x,r)|}\int_{B(x,\delta r)}\left(\int_{B(x,r)}\frac{d(y,z)}{|B(y,d(y,z))|}dy\right)|Xf^\star(z)|\,dz$$

(by Theorem 1.12) $\leq \frac{1}{|B(x,r)|}\int_{B(x,\delta r)}\left(\int_{B(z,(1+\delta)r)}\frac{d(y,z)}{|B(z,d(y,z))|}dy\right)|Xf^\star(z)|\,dz$

$$\leq C\,(1+\delta)\left(\frac{r}{|B(x,\delta r)|}\int_{B(x,\delta r)}|Xf^\star(z)|\,dz\right)$$

(by Proposition 1.15) $\leq C\,\dfrac{1+\delta}{\delta^\alpha}\,r^{1-\alpha}\left(\int_{B(x,\delta r)}|Xf^\star(y)|\dfrac{d(x,y)^\alpha}{|B(x,d(x,y))|}\,dy\right)$

$$= C\,r^{1-\alpha}\,I_\alpha(|Xf^\star|)(x).$$

This completes the proof. □

7.2. Fine mapping properties of fractional integration on metric spaces

The mapping properties of the operator I_α between Lebesgue spaces (or weighted Lebesgue spaces), are nowadays well-known, see the forthcoming book [**G02**]. Such properties are, however, of little or no use for our purposes, since we are facing the situation of two different measure spaces, with the target being relative to an upper (non-doubling) Ahlfors measure. The new ideas needed to tackle this problem originated in the Euclidean setting in the paper by D. Adams [**A71**] cited in the introduction. In what follows, we will need the following result from [**DGN98**] which generalizes the main result in [**A71**], and provides a ad hoc, more sophisticated version, of the classical fractional integration theorem.

THEOREM 7.7. *Let $U \subset \mathbb{R}^n$ be a given bounded set, with characteristic local parameters C_1, R_o, and Q. For $0 < \alpha < Q$, let $1 \leq p < \frac{Q}{\alpha}$. Suppose that for $0 \leq t < p$, μ is an upper t-Ahlfors measure. Under these assumptions, we have for any $B = B(x_o, R)$, $x_o \in U$, $0 < R \leq R_o$,*

$$I_\alpha : L^p(B, dx) \to L^{q,\infty}(B, d\mu),$$

with

$$q = p\,\frac{Q - \alpha t}{Q - \alpha p} > p.$$

Furthermore, there exists $C = C(C_1, \alpha, p, t, M) > 0$ such that, for any $f \in L^p(B(x_o, R), dx)$, and $\lambda > 0$,

$$\mu(\{x \in B(x_o, R)\,|\,I_\alpha f(x) > \lambda\}) \leq \frac{C}{\lambda^q}\left(\frac{R}{|B(x_o, R)|^{\frac{1}{Q}}}\right)^{\frac{Q(\alpha p - t)}{p(Q - t)}}\left(\int_{B(x_o, R)}|f|^p\,dx\right)^{\frac{q}{p}}.$$

7.3. Differentiation with respect to an upper Ahlfors measure

The main objective of this chapter is to prove the following qualitative version of trace theorem. We emphasize that the conclusion of Theorem 7.8 is relative to the measure μ, and not to the Lebesgue measure dx. We observe that, thanks to (1.22), the Lebesgue differentiation theorem for a space of homogeneous type in [**C76**] applies. From the latter we conclude that $f = f^\star \ dx - a.e.$, but the equivalence of f and $f^\star$ with respect to a different measure μ is a delicate matter (and, in fact, such equivalence cannot possibly subsist in general). Theorem 7.8, however, states that when μ is an upper Ahlfors measure, then $f = f^\star$ μ-a.e. This result will enable us to work with $f^\star$, instead of f, in all situations involving μ.

THEOREM 7.8. *Let $U \subset \mathbb{R}^n$ be a bounded set, with local parameters C_1, R_o, and Q. Given $1 \leq p < Q$, let μ be an upper s-Ahlfors measure for $0 \leq s < p$. Under these hypothesis, if $f \in \mathcal{L}^{1,p}(B(x_o, \delta r), dx)$, then*

$$(7.7) \qquad \lim_{t \to 0} \frac{1}{|B(x,t)|} \int_{B(x,t)} |f^\star(x) - f^\star(y)| \, dy = 0,$$

for μ-a.e. point $x \in B(x_o, \delta r)$.

Proof. We fix a number $t > 0$ such that $s < t < p$, and set $\alpha = s/t$, so that $0 < \alpha < 1$, and $s = \alpha t$. Let $Leb(f^\star)$ be the set of Lebesgue points of $f^\star$ with respect to Lebesgue measure dx, i.e.,

$$Leb(f^\star) = \left\{ x \in B(x_o, \delta r) \mid \lim_{t \to 0} \frac{1}{|B(x,t)|} \int_{B(x,t)} |f^\star(x) - f^\star(y)| \, dy = 0 \right\}.$$

A direct consequence of Theorem 7.6 is that

$$\{x \in B(x_o, \delta r) \mid I_\alpha(|Xf^\star|)(x) < \infty\} \subset Leb(f^\star),$$

or, equivalently,

$$B(x_o, \delta r) \setminus Leb(f^\star) \subset \{x \in B(x_o, \delta R) \mid I_\alpha(|Xf^\star|)(x) = \infty\}.$$

From this inclusion it is clear that, if we can prove that

$$(7.8) \qquad \mu(\{x \in B(x_o, \delta r) \mid I_\alpha(|Xf^\star|)(x) = \infty\}) = 0,$$

then we would also have $\mu(B(x_o, \delta r) \setminus Leb(f^\star)) = 0$. Therefore, (7.7) would hold at μ-a.e. point $x \in B(x_o, \delta r)$, thus completing the proof. To establish (7.8) we observe

$$E \stackrel{def}{=} \{x \in B(x_o, \delta r) \mid I_\alpha(|Xf^\star|)(x) = \infty\} \subset \bigcap_{\lambda > 0} \{x \in B(x_o, \delta r) \mid I_\alpha(|Xf^\star|)(x) > \lambda\}.$$

Our choice of α, and the assumption that μ is an upper s-Ahlfors measure, guarantee that we can apply Theorem 7.7, obtaining

$$\mu(E) \leq \mu(\{x \in B(x_o, \delta r) \mid I_\alpha(|Xf^\star|)(x) > \lambda\})$$

$$\leq \frac{C}{\lambda^q} \left(\frac{\delta r}{|B(x_o, \delta r)|^{\frac{1}{Q}}} \right)^{\frac{Q(\alpha p - t)}{p(Q-t)}} \left(\int_{B(x_o, \delta r)} |Xf^\star|^p \, dx \right)^{\frac{q}{p}},$$

for every $\lambda > 0$. Letting $\lambda \to \infty$ in the latter inequality, we infer (7.8). $\square$

An important consequence of Theorem 7.8 is that, if we identify f with $f^\star$, then f is defined everywhere with respect to μ. In particular, if $F \subset B(x_o, R)$ is such that $|F| = 0$, but $\mu(F) > 0$ (this is the case, for instance, when μ is the perimeter measure (4.7) supported on a lower dimensional manifold), then we can define the trace of f on F to be the pointwise restriction of $f^\star$ to F. Such restriction is then defined $\mu-a.e.$ on F. Using the results in [**DGN98**] a meaningful notion of trace on $F = supp\ \mu$ for functions in a Sobolev space $\mathcal{L}^{1,p}(B(x_o, \delta r), dx)$ can be obtained. However, when dealing with traces on the boundary of an open set Ω in such a general setting, one needs the Sobolev extension theorem (Theorem 9.4). Since such extension operators are not unique, to prove that the trace is independent of the choice of the extended function, Theorem 7.8 is required. We refer the reader to chapter 10 for details.

7.4. Upper Ahlfors measures and Hausdorff measure

We mention that a different form of Theorem 7.8 was established in [**HK98**] for an L^p Sobolev function f on a metric space. There, the authors proved that for any $k < p$ one has
$$\mathcal{H}^{Q-k}\left(\{x \mid M_\alpha g(x) = \infty\}\right) = 0,$$
where M_α is the Hardy-Littlewood fractional maximal operator, g is the "generalized gradient" of f, $\mathcal{H}^{Q-k}$ is the standard metric $(Q-k)$-Hausdorff measure, and $Q = \log_2 C_1$ is the homogeneous dimension of the metric space associated with the doubling constant C_1. Our Theorem 7.8 cannot be inferred from this result in [**HK98**] since it may not possible to control the measure μ from above by $\mathcal{H}^{Q-k}$. Instead, as the following result proves, if μ is a lower s-Ahlfors measure, then it is possible to control μ from below in terms of $\mathcal{H}^{Q-s}$. As a consequence, when μ is an s-Ahlfors measure, Theorem 7.8 implies the version found in [**HK98**].

DEFINITION 7.9. Given a set $E \subset \mathbb{R}^n$, we let
$$\mathcal{H}^s_\lambda(E) = inf\left\{\sum_{i=1}^\infty r_i^s \,\Big|\, E \subset \bigcup_{i=1}^\infty B(x_i, r_i), \quad r_i < \lambda\right\},$$
and
$$\mathcal{H}^s(E) = \lim_{\lambda \to 0} \mathcal{H}^s_\lambda(E).$$
We call $\mathcal{H}^s$ the s-dimensional *metric Hausdorff measure*.

For the properties of Hausdorff measures in metric spaces we refer the reader to [**Fe69**], [**M95**]. The following is a particular case (our $d(x, y)$ is an actual distance) of the Vitali type covering lemma for spaces of homogeneous type, see e.g., [**CW71**, p.69].

LEMMA 7.10. *Let $E \subset U$ be given and $\mathcal{G} = \{B(x, r(x))\}$ be any covering of E such that*
$$D = sup\{r(x) \mid B(x, r(x)) \in \mathcal{G}\} < \infty.$$
Then there is a countable sub-family $\mathcal{F} \subset \mathcal{G}$ of pairwise disjoint elements such that
$$E \subset \bigcup_{B \in \mathcal{F}} \hat{B}.$$

In the above, we have used $\hat{B}$ to denote the ball having the same center but with five times the radius of B.

THEOREM 7.11. *Let $U \subset \mathbb{R}^n$ be a bounded set with local parameters C_1, R_o, and Q. If for some $0 < s \leq Q$, μ is a lower s-Ahlfors measure, then there exists a constant $C = C(C_1, R_o, s, M) > 0$ such that, for every Borel set $E \subset\subset U$, one has*

$$\mu(E) \geq \frac{C}{M} \mathcal{H}^{Q-s}(E) .$$

Proof. We assume $\mu(E) < \infty$, otherwise, there is nothing to prove. Given $\epsilon > 0$, we choose an open set $A \supset E$ such that $\mu(A) < \mu(E) + \epsilon$. This is possible since μ is a Borel measure. We let

$$\mathcal{G}_\epsilon = \left\{ B(x,r) \,|\, x \in E, 0 < r < \frac{\epsilon}{2}, \text{ and } B(x,r) \subset A, \right\}.$$

Clearly, $\mathcal{G}_\epsilon$ covers E and thus, by Lemma 7.10, there exists a sequence of pairwise disjoint balls $\{B(x_i, r_i)\}_{i \in \mathbb{N}}$, all contained in A, such that

$$E \subset \bigcup_{i=1}^\infty B(x_i, 5r_i).$$

Definition 7.9, and the assumption on μ, give

$$\mathcal{H}^{Q-s}_{5\epsilon}(E) \leq \sum_{i=1}^\infty (5r_i)^{Q-s}$$

$$(\text{by } (1.24)) \leq C(C_1, R_o, s) \sum_{i=1}^\infty \frac{|B(x_i, r_i)|}{r_i^s}$$

$$\leq C\,M \sum_{i=1}^\infty \mu(B(x_i, r_i))$$

$$\leq C\,M\,\mu(A) \leq C\,M\,(\mu(E) + \epsilon).$$

Letting $\epsilon \to 0$ we reach the conclusion. $\square$

CHAPTER 8

Embedding a Sobolev space into a Besov space with respect to an upper Ahlfors measure

The main objective of this chapter is to prove that, under suitable conditions on s, given an upper s-Ahlfors measure μ, a function f in the Sobolev space $\mathcal{L}^{1,p}(B, dx)$ admits a *trace* $\tilde{f}$ on $F = \operatorname{supp} \mu$ which belongs to the sharp Besov space $B^p_{1-s/p}(F, d\mu)$. Furthermore, the trace operator $Tr : \mathcal{L}^{1,p}(B, dx) \to B^p_{1-s/p}(F, d\mu)$ is a bounded linear operator. For the precise statement, we refer the reader to Theorem 8.6. The proof of the latter is very delicate and involves a substantial amount of work. We begin with some preparatory results.

8.1. Some results from harmonic analysis

In the proof of Theorem 9.1 we will need the following adaption to a space of homogeneous type (which, for convenience, we take to be $(\mathbb{R}^n, d, dx)$, where $d = d(x, y)$ is the CC distance associated to X) of a classical result of Whitney. We refer the reader to [**St93**] for the proof of Theorem 8.1.

THEOREM 8.1 (**Whitney decomposition**). *Let $F \subset \mathbb{R}^n$ be a closed subset. There exists a family of balls $\mathcal{F} = \{B(p_i, r_i) \mid i = 1, 2, ...\}$ such that:*
  (i) $\mathbb{R}^n \setminus F = \bigcup_{B \in \mathcal{F}} B$;
  (ii) *if $i \neq j$, then $B(p_i, \frac{r_i}{2}) \cap B(p_j, \frac{r_j}{2}) = \varnothing$;*
  (iii) $\frac{7}{2} r(B_j) \leq d(B_j, F) \leq 5 r(B_j)$;
  (iv) *If $6B_i \cap 6B_j \neq \varnothing$ then $\frac{1}{24} r(B_j) \leq r(B_i) \leq 24 r(B_j)$;*
  (v) *$\mathcal{F}$ is locally finite. That is, for any $\alpha > 1$ there exists $N = N(\alpha)$, depending on the doubling constant in (1.22), such that if $B_{i_o} \in \mathcal{F}$ is a fixed ball, then the number of balls $B_j \in \mathcal{F}$, such that $\alpha B_j \cap \alpha B_{i_o} \neq \varnothing$, is less than or equal to N.*

In the above, we have let $B_j = B(p_j, r_j)$, $r_j = r(B_j)$, and $\alpha B_j = B(p_j, \alpha r_j)$. We next recall a generalization of the classical Calderón-Zygmund decomposition for spaces of homogeneous type established in [**Chr90**], see also [**St93**].

THEOREM 8.2. *Let U and R_o be as in Theorem 1.12, and let $\Omega \subset\subset U$ be open with $\operatorname{diam}(\Omega) < R_o/2$. There is constant $c > 0$, depending only on the characteristic local parameters of U, such that for every $f \in L^1(\Omega)$, $t > 0$, $C^* > 1$, there exist a decomposition of f,*

$$f = g + \sum_j b_j,$$

and a sequence of balls $\{B_j\} = \{B(p_j, r_j)\}$, such that

(i) $|g(x)| \leq c\, t$, for dx a.e. $x \in \Omega$;
(ii) $\operatorname{supp} b_j \subset B_j \subset C^* B_j \subset \Omega$ and, moreover,

$$\int b_j(y)\, dy = 0, \quad \text{and} \quad \int |b_j(y)|\, dy \leq c\, t\, |B_j|;$$

(iii) there exists an integer N such that each point in Ω belongs to at most N balls of the family $\{B_j\}$;
(iv) $\Omega \subset \bigcup_j 2C^* B_j$, and $\sum_j |B_j| \leq \frac{c}{t} \int |f(y)|\, dy$.

Note that since Theorem 1.12 is local in nature, we need to correspondingly localize Theorem 8.2 with respect to the global statements in [**Chr90**], [**St93**]. The relative details are standard and we omit them. The only difference with respect to the presentation in [**St93**] being that one needs to work with the following maximal function

$$Mf(x) = \sup_{x \in B} \frac{1}{|B \cap \Omega|} \int_{B \cap \Omega} |f(y)|\, dy.$$

It is worth pointing out that, since the supports of the b_j's are obtained from the Whitney decomposition in Theorem 8.1, it is possible to force $C^* B_j$ to be contained in Ω, see [**St93**, p.15]. When in chapter 8.3 we will use Theorem 8.2 in the proof of Lemma 8.5, we will have to make sure that the radii and centers of the balls B_j fall within the range of the parameters of Theorem 1.12.

8.2. Two simple growth-estimates

In the sequel we record two elementary, but useful, estimates concerning upper Ahlfors measures. Since their proof is elementary, we omit it altogether.

LEMMA 8.3. *Assume that ν is an upper τ-Ahlfors measure, with $\tau \geq 0$. Fix a bounded set $U \subset \mathbb{R}^n$, and let $\alpha > \tau$. There exists $C = C(C_1, \tau, M, \alpha) > 0$ such that for any $x \in U$, and $0 < r \leq R_o/2$*

$$\int_{B(x,r)} \frac{d(x,z)^\alpha}{|B(x, d(x,z))|}\, d\nu(z) \leq C\, r^{\alpha - \tau}.$$

When $d\nu = dx$, the Lebesgue measure on $\mathbb{R}^n$, we will take $\tau = 0$ in Lemma 8.3.

LEMMA 8.4. *Under the same assumptions of Lemma 8.3, let $\alpha < \tau$. There exists $C = C(C_1, \tau, M, \alpha) > 0$ such that for any $x \in U$, and $0 < r \leq R < R_o$ one has*

$$\int_{\{r < d(x,z) < R\}} \frac{d(x,z)^\alpha}{|B(x, d(x,z))|}\, d\nu(z) \leq C\, r^{\alpha - \tau}.$$

8.3. A key continuity estimate for a singular integral

The next lemma will be important in the proof of Theorem 8.6. Let $U \subset \mathbb{R}^n$ be a bounded open set, with characteristic local parameters C_1, R_o, and local homogeneous dimension Q. For some given numbers $\alpha, \gamma, \rho > 0$. For $x, z \in U$ with $x \neq z$, we define

$$K_\rho(x,z) = \begin{cases} \dfrac{d(x,z)^\alpha}{\Lambda(x,d(x,z))} & \text{if } 0 < d(x,z) < \rho, \\[2mm] \dfrac{\rho^{1+\gamma} d(x,z)^{\alpha-1-\gamma}}{\Lambda(x,d(x,z))} & \text{if } \rho \leq d(x,z), \end{cases}$$

where $\Lambda(x,r)$ represents the Nagel-Stein-Wainger polynomial in Definition 1.11 and in Theorem 1.12.

LEMMA 8.5. *Assume that μ is an upper s-Ahlfors measure with $0 < s \leq Q$. Let $\alpha < 1+s$. There exists $C = C(C_1, s, M, \alpha, \gamma) > 0$ such that for every $y, z \in U$ with $0 < d(y,z) < R_o/8$, $\rho < R_o/4$, one has*

$$\int_{\{\frac{R_o}{4} > d(x,z) > 2d(y,z)\}} |K_\rho(x,z) - K_\rho(x,y)|\, d\mu(x) \leq C\, d(y,z)^{\alpha-s}.$$

Proof. Our main objective is to prove that for every $x, y, z \in U$, such that

(8.1) $$2\, d(y,z) < d(x,z) < \frac{R_o}{4},$$

the following estimate holds

(8.2) $$|K_\rho(x,z) - K_\rho(x,y)| \leq C\, d(y,z)\, \frac{d(x,z)^{\alpha-1}}{|B(z,d(x,z))|}.$$

Once (8.2) is available we can easily reach the conclusion of the lemma as follows.

$$\int_{\{\frac{R_o}{4} > d(x,z) > 2d(y,z)\}} |K_\rho(x,z) - K_\rho(x,y)|\, d\mu(x)$$

$$\leq C\, d(y,z) \int_{\{\frac{R_o}{4} > d(x,z) > 2d(y,z)\}} \frac{d(x,z)^{\alpha-1}}{|B(z,d(x,z))|}\, d\mu(x)$$

$$\leq C\, d(y,z)^{\alpha-s},$$

where in the second to the last inequality we have used (8.2), whereas in the last we have exploited Lemma 8.4. This is possible since $\alpha - 1 < s$. In order to prove (8.2) we begin by making the simple observation that the inequality in the left-hand side of (8.1) implies

(8.3) $$\frac{1}{2} d(x,z) < d(x,y) < \frac{3}{2} d(x,z).$$

We now distinguish four cases:

(8.4) $$d(x,z),\ d(x,y) < \rho;$$

(8.5) $$d(x,z) < \rho, \qquad \rho \leq d(x,y) \leq \frac{R_o}{4};$$

we set $\sigma = 8(4\delta + 1)$. Fix $B_o = B(x_o, R) \subset U$, with $0 < R < \frac{R_o}{2\sigma}$, and let $f \in \mathcal{L}^{1,p}(\sigma B_o, dx)$. To prove the theorem we will show that

$$(8.14) \quad \|f\|_{L^p(B_o, d\mu)} \le \frac{C}{R^{\frac{s}{p}}} \left\{ \left(\int_{\sigma B_o} |f(y)|^p \, dy \right)^{\frac{1}{p}} + R \left(\int_{\sigma B_o} |Xf(y)|^p \, dy \right)^{\frac{1}{p}} \right\},$$

and

$$(8.15) \quad \mathcal{N}^p_{1-s/p}(f, F, d\mu) \le C \, \|Xf\|_{L^p(\sigma B_o, dx)}.$$

We first establish (8.14). Choose α such that $s/p < \alpha < 1$. This is possible since we assume $s < p$. Theorem 7.2 and (1.7) allow to obtain

$$\int_{B_o} |f(x)|^p \, d\mu(x)$$

$$\le C_p \int_{B_o} |f_{B_o}|^p \, d\mu(x) + C_p \int_{B_o} |f(x) - f_{B_o}|^p \, d\mu(x)$$

$$\le C_p \, \mu(B_o) |f_{B_o}|^p + C_p \int_{B_o} \left(\int_{\sigma B_o} \frac{d(x,y)}{|B(x, d(x,y))|} |Xf(y)| \, dy \right)^p d\mu(x)$$

$$\le C \, R^{-s} \int_{B_o} |f(y)|^p \, dy$$

$$+ C \int_{B_o} \left(\int_{\sigma B_o} \frac{d(x,y)^\alpha}{|B(x, d(x,y))|^{\frac{1}{p}}} \frac{d(x,y)^{1-\alpha}}{|B(x, d(x,y))|^{\frac{1}{p'}}} |Xf(y)| \, dy \right)^p d\mu(x)$$

$$\le C \, R^{-s} \int_{B_o} |f(y)|^p \, dy$$

$$+ C \int_{B_o} \left(\int_{\sigma B_o} \frac{d(x,y)^{\alpha p}}{|B(x, d(x,y))|} |Xf(y)|^p \, dy \right) \left(\int_{\sigma B_o} \frac{d(x,y)^{(1-\alpha)p'}}{|B(x, d(x,y))|} \, dy \right)^{\frac{p}{p'}} d\mu(x)$$

$$\le C \, R^{-s} \int_{B_o} |f(y)|^p \, dy + C \, R^{(1-\alpha)p} \int_{\sigma B_o} \left(\int_{B_o} \frac{d(x,y)^{\alpha p}}{|B(x, d(x,y))|} d\mu(x) \right) |Xf(y)|^p \, dy$$

$$\le C \, R^{-s} \int_{B_o} |f(y)|^p \, dy + C \, R^{(1-\alpha)p} \, R^{\alpha p - s} \int_{\sigma B_o} |Xf(y)|^p \, dy,$$

where in the first and second to the last inequality we have applied Lemma 8.3. We conclude that (8.14) holds. We next establish (8.15). For $x, y \in B_o$ we denote by B a ball containing x and y with radius $2d(x,y)$. Theorem 7.2 gives

$$(8.16) \quad |f(x) - f(y)| \le |f(x) - f_B| + |f(y) - f_B|$$

$$\le C \int_{\delta B} \frac{d(x,z)}{|B(x, d(x,z))|} |Xf(z)| \, dz + C \int_{\delta B} \frac{d(y,z)}{|B(y, d(y,z))|} |Xf(z)| \, dz.$$

The radius of the ball δB in (8.16) is less than $4\delta R$, therefore by our choice of σ we have $\delta B \subset \frac{\sigma}{8} B_o$. Since the role of x and y in (8.16) can be reversed, it suffices to consider only one of the two integrals in the right-hand side. With $\beta = 1 - s/p$ we obtain from (8.16)

(8.17)
$$\begin{aligned}\mathcal{N}_\beta^p&(f,F,d\mu)^p\\ &\leq C\int_{B_o}\int_{B_o}\frac{d(x,y)^{s-\beta p}}{|B(x,d(x,y))|}\left(\int_{\delta B}\frac{d(x,z)}{|B(x,d(x,z))|}|Xf(z)|\,dz\right)^p d\mu(x)\,d\mu(y)\\ &\leq C\int_{B_o}\int_{B_o}\frac{d(x,y)^{s-\beta p}}{|B(x,d(x,y))|}\\ &\qquad\left(\int_{B(x,2(1+\delta)d(x,y))}\frac{d(x,z)}{|B(x,d(x,z))|}|Xf(z)|\,dz\right)^p d\mu(x)\,d\mu(y).\end{aligned}$$

For a fixed $x \in B_o$ we have from (1.7) and (1.22)

(8.18)
$$\begin{aligned}\int_{B_o}&\frac{d(x,y)^{s-\beta p}}{|B(x,d(x,y))|}\left(\int_{B(x,2(1+\delta)d(x,y))}\frac{d(x,z)}{|B(x,d(x,z))|}|Xf(z)|\,dz\right)^p d\mu(y)\\ &\leq \sum_{i=0}^{\infty}\int_{\{y|2^{-i}R<d(x,y)\leq\,2^{-i+1}R\}}\frac{d(x,y)^{s-\beta p}}{|B(x,d(x,y))|}\\ &\qquad\left(\int_{B(x,2(1+\delta)d(x,y))}\frac{d(x,z)}{|B(x,d(x,z))|}|Xf(z)|\,dz\right)^p d\mu(y)\\ &\leq 2^s\,M\,R^{s-\beta p}\sum_{i=0}^{\infty}\frac{(2^{-i})^{s-\beta p}}{|B(x,2^{-i}R)|}\frac{|B(x,2^{-i+1}R)|}{(2^{-i+1}R)^s}\\ &\qquad\left(\int_{\{d(x,z)<4(1+\delta)2^{-i}R\}}\frac{d(x,z)}{|B(x,d(x,z))|}|Xf(z)|\,dz\right)^p\\ &\leq C\,R^{-\beta p}\sum_{i=0}^{\infty}2^{i\beta p}\left(\int_{\{d(x,z)<4(1+\delta)2^{-i}R\}}\frac{d(x,z)}{|B(x,d(x,z))|}|Xf(z)|\,dz\right)^p.\end{aligned}$$

We let $F(r) = \left(\int_{\{d(x,z)<\frac{\sigma}{4}r\}}\frac{d(x,z)}{|B(x,d(x,z))|}|Xf(z)|\,dz\right)^p$. Since F is increasing in r, one has

(8.19)
$$\begin{aligned}\int_0^R r^{-1-\beta p}F(r)\,dr &= \sum_{i=0}^{\infty}\int_{\{2^{-i-1}R<r\leq 2^{-i}R\}}r^{-1-\beta p}F(r)\,dr\\ &\geq R^{-\beta p}\sum_{i=0}^{\infty}2^{i\beta p}F(2^{-i-1}R)\\ &= \frac{R^{-\beta p}}{2}\sum_{i=0}^{\infty}2^{i\beta p}\left(\int_{\{d(x,z)<\frac{\sigma}{8}2^{-i}R\}}\frac{d(x,z)}{|B(x,d(x,z))|}|Xf(z)|\,dz\right)^p.\end{aligned}$$

Using (8.19) in (8.18), and (8.18) in (8.17), we obtain

(8.20)
$$\mathcal{N}_\beta^p(f,B_o)^p \le C \int_{B_o} \int_0^R r^{-1-\beta p} \left(\int_{\{d(x,z)<\frac{\sigma}{4}r\}} \frac{d(x,z)}{|B(x,d(x,z))|} |Xf(z)|\,dz \right)^p dr\,d\mu(x)$$
$$\le C \int_{B_o \times [0,R)} T(|Xf|)(x,r)^p\,d\mu_1(x,r),$$

where we have let

(8.21) $$d\mu_1(x,r) = r^{-1+s}\,dr\,d\mu(x),$$

and defined

(8.22) $$Th(x,r) = \frac{1}{r}\int_{\{d(x,z)<\frac{\sigma}{4}R\}} \tilde{K}_{\frac{\sigma}{4}r}(x,z)|h(z)|\,dz,$$

with

$$\tilde{K}_\rho(x,z) = \begin{cases} \dfrac{d(x,z)}{|B(x,d(x,z))|} & \text{if } 0 < d(x,z) < \rho, \\ \dfrac{\rho^{1+\frac{1+\gamma}{p}}d(x,z)^{-\frac{1+\gamma}{p}}}{|B(x,d(x,z))|} & \text{if } \rho \le d(x,z). \end{cases}$$

The parameter $\gamma > 0$ in the definition of the kernel $\tilde{K}_\rho(x,z)$ is fixed so that $s < \gamma$. The role of the additional term in the definition of $\tilde{K}_\rho$ will become clear once we reach (8.36) below. Essentially, the purpose of such term is to smoothly cut-off K_ρ, rather than simply truncate it, for $d(x,z) \ge \rho$. Keeping in mind (8.20) we see that in order to establish (8.15) it suffices to prove that the sub-linear operator T maps boundedly $L^p(\sigma B_o, dx)$ into $L^p(B_o \times [0,R), d\mu_1)$. In view of Marcinkiewicz interpolation theorem, it is enough to show that T is of weak type (p,p) for all $p > 1$. We will thus prove

(8.23) $$\mu_1(\{(x,r) \in B_o \times [0,R) \mid Th(x,r)^p > \lambda^p\}) \le C\left(\frac{\|h\|_{L^p(\sigma B_o)}}{\lambda}\right)^p.$$

We now fix $a > 0$ in the following way

(8.24) $$\frac{s}{p} < a < \min\left\{1, \frac{s+1}{p}\right\}.$$

We notice that, since $s < \gamma$, in view of (8.24) the number a also satisfies

(8.25) $$a < \frac{\gamma+1}{p}.$$

Let p' be such that $\frac{1}{p} + \frac{1}{p'} = 1$. One has

$$Th(x,r)^p \le C_p\,r^{-p}\left(\int_{\{d(x,z)<\frac{\sigma}{4}r\}} \frac{d(x,z)^a}{|B(x,d(x,z))|^{\frac{1}{p}}} \frac{d(x,z)^{(1-a)}}{|B(x,d(x,z))|^{\frac{1}{p'}}}|h(z)|\,dz\right)^p$$
$$+ C(p,\sigma)\,r^{-p}\left(\int_{\{\frac{\sigma}{4}R>d(x,z)\ge\frac{\sigma}{4}r\}} \frac{r^{\frac{1+\gamma}{p}}d(x,z)^{a-\frac{1+\gamma}{p}}}{|B(x,d(x,z))|^{\frac{1}{p}}}\frac{r\,d(x,z)^{-a}}{|B(x,d(x,z))|^{\frac{1}{p'}}}|h(z)|\,dz\right)^p$$

$$\leq C\, r^{-p} \left(\int_{\{d(x,z)<\frac{\sigma}{4}r\}} \frac{d(x,z)^{ap}}{|B(x,d(x,z))|} |h(z)|^p\, dz \right)$$

$$\left(\int_{\{d(x,z)<\frac{\sigma}{4}r\}} \frac{d(x,z)^{(1-a)p'}}{|B(x,d(x,z))|}\, dz \right)^{\frac{p}{p'}}$$

$$+ C\, r^{-p} \left(\int_{\{\frac{\sigma}{4}R>d(x,z)\geq \frac{\sigma}{4}r\}} \frac{r^{1+\gamma}\, d(x,z)^{ap-1-\gamma}}{|B(x,d(x,z))|} |h(z)|^p\, dz \right)$$

$$\left(\int_{\{\frac{\sigma}{4}R>d(x,z)\geq \frac{\sigma}{4}r\}} \frac{r^{p'}\, d(x,z)^{-ap'}}{|B(x,d(x,z))|}\, dz \right)^{\frac{p}{p'}}$$

$$\leq C\, r^{-ap} \left\{ \int_{\{d(x,z)<\frac{\sigma}{4}r\}} \frac{d(x,z)^{ap}}{|B(x,d(x,z))|} |h(z)|^p\, dz \right.$$

$$\left. + \int_{\{\frac{\sigma}{4}R>d(x,z)\geq \frac{\sigma}{4}r\}} \frac{r^{1+\gamma}\, d(x,z)^{ap-1-\gamma}}{|B(x,d(x,z))|} |h(z)|^p\, dz \right\},$$

where in the last inequality we have used Lemmas 8.3 and 8.4, with $d\nu = dx$, and $\tau = 0$. We stress that Lemma 8.3 can be applied since by our choice $a < 1$, see (8.24). On the other hand, it is possible to implement Lemma 8.4 since, thanks to (8.25), we have $ap - 1 - \gamma < 0$. We thus conclude

(8.26)
$$Th(x,r)^p \leq C\, r^{-ap} \left\{ \int_{\{d(x,z)<\frac{\sigma}{4}R\}} K_{\frac{\sigma}{4}r}(x,z)|h(z)|^p\, dz + \int_{\sigma B_o} K_{\frac{\sigma}{4}r}(x,z)|h(z)|^p\, dz \right\},$$

where K_ρ is the kernel introduced in Lemma 8.5 corresponding to the choice $\alpha = ap$. We now apply Theorem 8.2 to $|h|^p \in L^1(\sigma B_o)$, with $C^* = 2\sigma$, at level $t = c_o \lambda^p$. The value of c_o will be conveniently chosen subsequently. Let $|h(z)|^p = g(z) + \sum_j b_j(z)$ be the corresponding decomposition, with family of balls $\{B_j\}_{j\in\mathbb{N}}$, $B_j = B(p_j, r_j)$. From (8.26) we obtain

(8.27)
$$Th(x,r)^p \leq H_1(x,r) + H_2(x,r),$$

where

$$H_1(x,r) = C\, r^{-ap} \int_{\sigma B_o} K_{\frac{\sigma}{4}r}(x,z) g(z)\, dz$$

$$H_2(x,r) = C\, r^{-ap} \int_{\sigma B_o} K_{\frac{\sigma}{4}r}(x,z) \sum_j b_j(z)\, dz.$$

From (8.27) we obtain

$$\mu_1(\{(x,r) \in B_o \times [0,R)\,|\, Th(x,r)^p > \lambda^p\})$$

$$\leq \mu_1(\{(x,r) \in B_o \times [0,R)\,|\, H_1(x,r) > \frac{\lambda^p}{2}\})$$

$$+ \mu_1(\{(x,r) \in B_o \times [0,R)\,|\, H_2(x,r) > \frac{\lambda^p}{2}\}).$$

In order to prove (8.23) it will thus suffice to establish similar estimates for H_1 and H_2. In fact, we will soon prove that, for a suitable choice of the constant c_o, we have

(8.28) $$H_1(x,r) \leq \frac{\lambda^p}{2}.$$

Consequently,

$$\mu_1(\{(x,r) \in B_o \times [0,R) \mid H_1(x,r) > \frac{\lambda^p}{2}\}) = 0,$$

and therefore, to achieve (8.23), we only need to prove

(8.29) $$\mu_1(\{(x,r) \in B_o \times [0,R) \mid H_2(x,r) > \frac{\lambda^p}{2}\}) \leq C \left(\frac{\|h\|_{L^p(\sigma B_o)}}{\lambda}\right)^p.$$

To establish (8.28) we use (i) of Theorem 8.2, and again apply Lemmas 8.3, 8.4 with $d\nu$ equal to Lebesgue measure dx, and $\tau = 0$, obtaining

$$H_1(x,r) \leq C\, r^{-ap}\, c\, c_o\, \lambda^p \left\{ \int_{\{z \in \sigma B_o \mid d(x,z) < \frac{\sigma}{4}r\}} \frac{d(x,z)^{ap}}{|B(x,d(x,z))|}\, dz \right.$$
$$\left. + C(\sigma,\gamma) \int_{\{z \in \sigma B_o \mid \frac{\sigma}{4}r \leq d(x,z) < R_o\}} \frac{r^{1+\gamma} d(x,z)^{ap-1-\gamma}}{|B(x,d(x,z))|}\, dz \right\}$$
$$\leq \kappa\, c_o\, \lambda^p\, r^{-ap}\, (r^{ap} + r^{1+\gamma} r^{ap-1-\gamma}) = \kappa\, c_o\, \lambda^p.$$

Here $\kappa > 0$ is a constant which depends on the characteristic local parameters of U, and on p, s, γ, a. At this point, with the choice $c_o = \frac{1}{2\kappa}$, we see that (8.28) holds. We then turn to the more delicate estimate (8.29). We let

$$E_j = B(p_j, 2\sigma r_j) \times [0, r_j), \qquad E = \cup_j E_j \subset (\sigma B_o \times [0,R)).$$

Denoting by E^c the complement of E with respect to $\sigma B_o \times [0,R)$, one clearly obtains

(8.30) $$\{(x,r) \in B_o \times [0,R) \mid H_2(x,r) > \frac{\lambda^p}{2}\}$$
$$= \{(x,r) \in E \mid H_2(x,r) > \frac{\lambda^p}{2}\} \cup \{(x,r) \in E^c \mid H_2(x,r) > \frac{\lambda^p}{2}\}.$$

To estimate the μ_1 measure of the first set in the right-hand side of (8.30) we use the assumption (1.7), Theorem 1.12, and property (iv) in Theorem 8.2

(8.31) $$\mu_1(\{(x,r) \in E \mid H_2(x,r) > \frac{\lambda^p}{2}\}) \leq \mu_1(E) \leq \sum_j \mu_1(E_j)$$
$$= \sum_j \int_{E_j} d\mu_1(x,r) = \sum_j \int_{B(p_j, 2\sigma r_j)} \int_0^{r_j} r^{-1+s}\, dr\, d\mu(x)$$
$$= \frac{1}{s} \sum_j r_j^s \mu(B(p_j, 2\sigma r_j)) \leq C\, M \sum_j r_j^s \frac{|B(p_j, r_j)|}{r_j^s}$$
$$\leq \frac{CM}{\lambda^p} \int_{\sigma B_o} |h(z)|^p\, dz$$

8.4. THE MAIN THEOREM

As for the second set in the right-hand side of (8.30), observing that $E^c = (\cup E_j)^c \subset E_j^c$ for every $j \in \mathbb{N}$, one finds

(8.32)
$$\int_{E^c} |H_2(x,r)|\, d\mu_1(x,r) \leq C \sum_j \int_{E_j^c} r^{-ap} \left| \int_{\sigma B_o} K_{\frac{\sigma}{4}r}(x,z) b_j(z)\, dz \right| d\mu_1(x,r)$$
$$= C \sum_j (\mathcal{A}_j + \mathcal{B}_j),$$

where

$$\mathcal{A}_j = \int_{\{(x,r)\in B_o\times[0,R)\,|\,r_j\leq r\leq R\}} r^{-ap} \left| \int_{B(x_o,\sigma R)} K_{\frac{\sigma}{4}r}(x,z) b_j(z)\, dz \right| d\mu_1(x,r)$$

$$\mathcal{B}_j = \int_{\{(x,r)\in B_o\times[0,R)\,|\,0\leq r\leq r_j, d(x,m_j)\geq 2\sigma r_j\}} r^{-ap} \left| \int_{\sigma B_o} K_{\frac{\sigma}{4}r}(x,z) b_j(z)\, dz \right| d\mu_1(x,r).$$

We will prove that

(8.33)
$$\sum_j (\mathcal{A}_j + \mathcal{B}_j) \leq C \sum_j \int_{B(p_j,r_j)} |b_j(z)|\, dz.$$

Suppose we have achieved this. From (8.33), and from (ii) and (iv) of Theorem 8.2, we would conclude

$$\int_{E^c} |H_2(x,r)|\, d\mu_1(x,r) \leq C \int_{\sigma B_o} |h(z)|^p\, dz.$$

Finally, Chebychev inequality would give

$$\mu_1\left(\{(x,r) \in E^c \,|\, H_2(x,r) > \frac{\lambda^p}{2}\}\right)$$
$$\leq \frac{2}{\lambda^p} \int_{E^c} |H_2(x,r)|\, d\mu_1(x,r) \leq \frac{C}{\lambda^p} \int_{\sigma B_o} |h(z)|^p\, dz.$$

Together with (8.31) this would establish (8.29), and therefore (8.23), thus completing the proof of the theorem. We thus turn to proving (8.33). Observe that in the inner integral in the right-hand side of the formulas defining $\mathcal{A}_j, \mathcal{B}_j$, the integration really takes place on $\mathrm{supp}\, b_j \subset B(p_j, r_j)$. On the other hand, if $z \in B(p_j, r_j)$, then $B(p_j, 2\sigma r_j)^c \subset B(z, \sigma r_j)^c$. We can therefore estimate $\mathcal{B}_j$ as follows

(8.34)
$$\mathcal{B}_j \leq \int_{B(p_j,2\sigma r_j)^c} \int_0^{r_j} r^{-1+s-ap} \left(\int_{B(p_j,r_j)} K_{\frac{\sigma}{4}r}(x,z) |b_j(z)|\, dz \right) dr\, d\mu(x)$$
$$= \int_0^{r_j} r^{-1+s-ap} \int_{B(p_j,r_j)} |b_j(z)| \left(\int_{B(p_j,2\sigma r_j)^c} K_{\frac{\sigma}{4}r}(x,z)\, d\mu(x) \right) dz\, dr$$
$$\leq \int_0^{r_j} r^{-1+s-ap} \int_{B(p_j,r_j)} |b_j(z)| \left(\int_{B(z,2\sigma r_j)^c} K_{\frac{\sigma}{4}r}(x,z)\, d\mu(x) \right) dz\, dr$$

(8.35)
$$\leq C \int_0^{r_j} r^{-1+s-ap} \int_{B(p_j,r_j)} |b_j(z)|$$
$$\left(\int_{\{\sigma r_j \leq d(x,z) < R_o/4\}} \frac{r^{1+\gamma} \, d(x,z)^{ap-1-\gamma}}{|B(z,d(x,z))|} d\mu(x) \right) dz \, dr$$
$$\leq C \int_{B(p_j,r_j)} |b_j(z)| \int_0^{r_j} r^{-1+s-ap}$$
$$\left(\int_{\{\sigma r_j \leq d(x,z) < R_o/4\}} \frac{r \, d(x,z)^{ap-1}}{|B(z,d(x,z))|} d\mu(x) \right) dr \, dz$$
$$\leq C \, r_j^{ap-1-s} \int_{B(p_j,r_j)} |b_j(z)| \left(\int_0^{r_j} r^{s-ap} \, dr \right) dz$$
$$= C \int_{B(p_j,r_j)} |b_j(z)| \, dz,$$

where in the last inequality we have used Lemma 8.4 with $\nu = \mu$, and $\tau = s$. We note that this is possible since, thanks to (8.24), we have $ap - 1 < s$. This establishes half of (8.33). We are left with estimating $\mathcal{A}_j$. Since $\int_{B(x_o, \sigma R)} b_j(z) \, dz = 0$ we have

(8.36)
$$\mathcal{A}_j$$
$$= \int_{\{(x,r) \in B_o \times [0,R) \mid r_j \leq r \leq R\}} r^{-ap} \left| \int_{B(p_j,r_j)} \left(K_{\frac{\sigma}{4}r}(x,z) - K_{\frac{\sigma}{4}r}(x,p_j) \right) b_j(z) \, dz \right| d\mu_1(x,r)$$
$$\leq \int_{\{(x,r) \in B_o \times [0,R) \mid r_j \leq r \leq R\}} r^{-ap} \int_{B(p_j,r_j)} |K_{\frac{\sigma}{4}r}(x,z) - K_{\frac{\sigma}{4}r}(x,p_j)| \, |b_j(z)| \, dz \, d\mu_1(x,r)$$
$$= \int_{B(p_j,r_j)} |b_j(z)| \int_{r_j}^R r^{-1+s-ap} \left(\int_{B_o} |K_{\frac{\sigma}{4}r}(x,z) - K_{\frac{\sigma}{4}r}(x,p_j)| \, d\mu(x) \right) dr \, dz.$$

We analyze the innermost integral in (8.36) as follows.

(8.37)
$$\int_{B_o} |K_{\frac{\sigma}{4}r}(x,z) - K_{\frac{\sigma}{4}r}(x,p_j)| \, d\mu(x)$$
$$= \int_{\{x \in B_o \mid d(x,z) \leq 2d(p_j,z)\}} |K_{\frac{\sigma}{4}r}(x,z) - K_{\frac{\sigma}{4}r}(x,p_j)| \, d\mu(x)$$
$$+ \int_{\{x \in B_o \mid d(x,z) \geq 2d(p_j,z)\}} |K_{\frac{\sigma}{4}r}(x,z) - K_{\frac{\sigma}{4}r}(x,m_j)| \, d\mu(x)$$
$$\leq \int_{\{d(x,z) \leq 2d(p_j,z)\}} |K_{\frac{\sigma}{4}r}(x,z)| \, d\mu(x) + \int_{\{d(x,p_j) \leq 3d(p_j,z)\}} |K_{\frac{\sigma}{4}r}(x,p_j)| \, d\mu(x)$$
$$+ \int_{\{x \in B_o \mid d(x,z) \geq 2d(p_j,z)\}} |K_{\frac{\sigma}{4}r}(x,z) - K_{\frac{\sigma}{4}r}(x,p_j)| \, d\mu(x)$$

(8.38)
$$\leq \int_{\{d(x,z)\leq 2\ d(p_j,z)\}} \frac{d(x,z)^{ap}}{|B(x,d(x,z))|}\, d\mu(x)$$
$$+ \int_{\{d(x,p_j)\leq 3d(p_j,z)\}} \frac{d(x,p_j)^{ap}}{|B(p_j,d(x,p_j))|}\, d\mu(x) \;+\; C\, d(z,p_j)^{ap-s}$$
$$\leq C\, d(p_j,z)^{ap-s}.$$

In the derivation of (8.37) we have used in the order Lemma 8.5, Theorem 1.12, and Lemma 8.3. We stress that, in order to apply Lemma 8.5, we must have $0 < ap - s < 1$. This is guaranteed by (8.24). Inserting (8.37) in (8.36), in view of the assumption $0 < s < p$, we can conclude

$$\mathcal{A}_j \leq C \int_{B(p_j,r_j)} d(p_j,z)^{ap-s}\, |b_j(z)| \int_{r_j}^{R} r^{-1+s-ap}\, dr\, dz$$
$$\leq C \int_{B(p_j,r_j)} |b_j(z)|\, dz.$$

This establishes the remaining half of (8.33) and thereby completes the proof of the theorem. $\square$

REMARK 8.7. If one is only interested in the case $0 < \beta < 1 - s/p$, then the proof of Theorem 8.6 is much simpler and can be obtained as follows. Choosing $a > 0$ such that $s < ap < 1$, and proceeding from the first term in the right-hand side of (8.20), one has

$$\mathcal{N}_\beta^p(f, B_o, d\mu)^p$$
$$\leq C \int_{B_o} \int_0^R r^{-1-\beta p} \left(\int_{\{d(x,z)<\frac{\sigma}{4}r\}} \frac{d(x,z)}{|B(x,d(x,z))|} |Xf(z)|\, dz \right)^p dr\, d\mu(x)$$
$$\leq C \int_{B_o} \int_0^R r^{-1-\beta p} \left(\int_{\{d(x,z)<\frac{\sigma}{4}r\}} \frac{d(x,z)^{ap}}{|B(x,d(x,z))|} |Xf(z)|^p\, dz \right)$$
$$\times \left(\int_{\{d(x,z)<\frac{\sigma}{4}r\}} \frac{d(x,z)^{(1-a)p'}}{|B(x,d(x,z))|}\, dz \right)^{\frac{p}{p'}} dr\, d\mu(x)$$
$$\leq C \int_{B_o} \int_0^R r^{-1-\beta p}\, r^{(1-a)p} \left(\int_{\{d(x,z)<\frac{\sigma}{4}r\}} \frac{d(x,z)^{ap}}{|B(x,d(x,z))|} |Xf(z)|^p\, dz \right) dr\, d\mu(x)$$
$$\leq C \int_0^R r^{-1-\beta p+(1-a)p} \int_{B(x_o,\sigma R)} |Xf(z)|^p \left(\int_{\{d(x,z)<\frac{\sigma}{4}r\}} \frac{d(x,z)^{ap}}{|B(z,d(x,z))|}\, d\mu(x) \right) dr\, dz$$
$$\leq C \left(\int_0^R r^{-1-\beta p+(1-a)p}\, r^{ap-s}\, dr \right) \|Xf\|^p_{L^p(B(x_o,\sigma R), dx)}$$
$$= \frac{C}{p(1-s/p-\beta)} R^{p(1-s/p-\beta)} \|Xf\|^p_{L^p(B(x_o,\sigma R), dx)}.$$

In the above estimate we have used Lemma 8.3 twice, the former time with $d\nu = dx$, and $\tau = 0$, the latter with $d\nu = d\mu$, and $\tau = ap$. Such simple proof fails miserably in the end-point case $\beta = 1 - s/p$.

A situation of special interest arises when the set F in Theorem 8.6 is a portion of a $C^{1,1}$ hypersurface, and μ represents the perimeter measure introduced in Definition 4.7.

THEOREM 8.8. *Let $U \subset \mathbb{R}^n$ be a bounded set with characteristic local parameters C_1, R_o. There exists $\sigma = \sigma(X, U) > 0$ such that, given $p > 1$, $B_o = B(x_o, R) \subset U$, $0 < R < \frac{R_o}{2\sigma}$, $f \in \mathcal{L}^{1,p}(\sigma B_o, dx)$, and $F \subset B_o$, where $F \subset \partial\Omega$, and Ω is a bounded $C^{1,1}$ domain of type ≤ 2, one has*

$$\|f\|_{B^p_{1-\frac{1}{p}}(F, d\mu)} \leq C\|f\|_{\mathcal{L}^{1,p}(\sigma B_o, dx)}, \tag{8.39}$$

for some $C = C(X, U, p, \Omega) > 0$, where μ is the perimeter measure.

Proof. In view of Theorem 5.6 the perimeter measure of Ω is an upper 1-Ahlfors measure. We can therefore apply Theorem 8.6 to obtain the conclusion. □

CHAPTER 9

The extension theorem for a Besov space with respect to a lower Ahlfors measure

The main objective of this chapter is establishing the following result.

THEOREM 9.1 (**Extension theorem**). *Let $1 \leq p < \infty$, and μ be a lower s-Ahlfors measure in $\mathbb{R}^n$, with $0 < s < p$. When $p > n$ we require, in addition, that $s \leq \frac{n+p}{2}$. We assume that for an open set $\Omega \subset \mathbb{R}^n$, we have $F = \operatorname{supp} \mu$ be a compact subset of Ω, with $|F| = 0$. There exists a bounded linear mapping (an extension operator) $\mathcal{E} : B^p_{1-\frac{s}{p}}(F, d\mu) \to \mathcal{L}^{1,p}(\Omega, dx)$, such that*

(i) $\mathcal{E}f(x) = f(x)$ for μ a.e. $x \in F$, (ii) $\|\mathcal{E}f\|_{\mathcal{L}^{1,p}(\Omega,dx)} \leq C \|f\|_{B^p_{1-s/p}(F,d\mu)}$,

for some $C = C(X, p, s, M, \operatorname{dist}(F, \partial\Omega)) > 0$. Furthermore, $\mathcal{E}f$ is supported in a neighborhood of F.

We stress that while in the previous chapters the measure μ was assumed to be an upper Ahlfors measure, in the above theorem we need the control from below expressed by (1.8).

9.1. Some auxiliary results

The proof of Theorem 9.1 is based, among other things, on the extension theorem for the Sobolev spaces $\mathcal{L}^{1,p}$ established in [**GN98**]. We recall this basic result in Theorem 9.4 below. We will also need the following Whitney partition of unity on CC balls established in [**GN98**].

THEOREM 9.2. *Let $\mathcal{F}$ be the covering given by Theorem 8.1. There exists a partition of unity $\{\phi_j \,|\, j = 1, 2, ...\}$ subordinated to $\mathcal{F}$. That is, the ϕ_j satisfy*

(i) $\phi_j \in Lip_d(B_j^*)$, $\operatorname{supp} \phi_j \subset B_j^*$
(ii) $0 \leq \phi_j \leq 1$, $\sum_j \phi_j \equiv 1$ on $\bigcup_{B_j \in \mathcal{F}} B_j^* \supset \mathbb{R}^n \setminus F$,
(iii) $|X\phi_j| \leq \frac{c}{r(B_j)}$.

We next recall a notion from [**GN98**], which generalizes that of Euclidean (ϵ, δ)–domain introduced in [**Jo81**].

DEFINITION 9.3. *An open set $\Omega \subset \mathbb{R}^n$ is called an (ϵ, δ)-domain if there exist $0 < \delta \leq \infty$, $0 < \epsilon \leq 1$, such that for any pair of points $p, q \in \Omega$, if $d(p, q) \leq \delta$,* then one can find a continuous, rectifiable curve $\gamma : [0, T] \to \Omega$, for which $\gamma(0) = p$, $\gamma(T) = q$, and

(9.1) $\quad l(\gamma) \leq \dfrac{1}{\epsilon} d(p, q), \qquad d(z, \partial \Omega) \geq \epsilon \min \{d(p, z), d(z, q)\} \quad \text{for all } z \in \{\gamma\}.$

If $\Omega \subset \mathbb{R}^n$ is an open set, then *the radius* of Ω is defined as follows
(9.2)
$rad(\Omega) = \sup \{r > 0 \mid \partial B(p, s) \cap \Omega \neq \emptyset, \text{for every } p \in \Omega, \text{ and every } 0 \leq s < r\}.$
Note that if $\delta = \infty$, or Ω is connected, then $rad(\Omega) > 0$.

THEOREM 9.4. *Let $1 \leq p \leq \infty$. If $\Omega \subset \mathbb{R}^n$ is a bounded (ϵ, δ)-domain with $rad(\Omega) > 0$, then there exists a linear operator $\mathcal{S} : \mathcal{L}^{1,p}(\Omega, dx) \to \mathcal{L}^{1,p}(\mathbb{R}^n, dx)$ such that for some $C > 0$ one has for $f \in \mathcal{L}^{1,p}(\Omega, dx)$*

(i) $\mathcal{S}f(x) = f(x) \quad \text{for a.e. } x \in \Omega,$ \quad (ii) $\|\mathcal{S}f\|_{\mathcal{L}^{1,p}(\mathbb{R}^n, dx)} \leq C\|f\|_{\mathcal{L}^{1,p}(\Omega, dx)}.$

We have used here $\mathcal{S}$ to denote the extension operator for Sobolev spaces since we have reserved the symbol $\mathcal{E}$ to indicate the extension operator for Besov spaces. We are now ready to prove the main result of this chapter.

9.2. Proof of Theorem 9.1

We fix an open set Ω and a bounded set $U \subset \mathbb{R}^n$, such that $F \subset \Omega \subset\subset U$, and let C_1, R_o be the characteristic local parameters of U. Consider the Whitney decomposition $\mathcal{F}$ of $\mathbb{R}^n \setminus F$ as in Theorem 8.1. From the proof of the latter one can easily deduce the following. For any ball $B \in \mathcal{F}$ we have $6B \cap F \neq \emptyset$. We let $B^* = 6B$. There exist constants α_1, α_2 such that if $B, B_{i_o} \in \mathcal{F}$, and $B^* \cap B_{i_o} \neq \emptyset$, then

(9.3) $\qquad\qquad\qquad B_{i_o} \subset \alpha_1 B^* \subset \alpha_2 B_{i_o}.$

We let
$$\mathcal{F}' = \left\{ B \in \mathcal{F} \mid r(B) < \min\left\{ \dfrac{dist\, (F, \partial \Omega)}{30}, R_o \right\} \right\},$$
and notice that if $B \in \mathcal{F}'$, then $B^* \subset \Omega$. We define
(9.4)
$$\mathcal{E}f(x) = \begin{cases} f(x) & \text{if } x \in F, \\ \sum_{B_j \in \mathcal{F}} c_j \phi_j(x) & \text{if } x \in \Omega \setminus F, \end{cases} \quad \text{where } c_j = \dfrac{1}{\mu(\alpha_1 B_j^*)} \int_{\alpha_1 B_j^*} f(t)\, d\mu(t).$$

In (9.4), the collection of functions $\{\phi_j\}$ denotes the partition of unity in Theorem 9.2. It is clear from Theorem 8.1 that $\mu(\alpha_1 B_j^*) > 0$. Hence $\mathcal{E}$ is well defined, and it is supported in a neighborhood of F. We now proceed to prove that $\mathcal{E}$ is a bounded operator by showing there exists a constant $C > 0$ such that

(a) $\displaystyle\int_{\Omega \setminus F} |\mathcal{E}f(x)|^p\, dx \leq C \int_F |f(x)|^p\, d\mu(x),$

(b) $\displaystyle\int_{\Omega \setminus F} |X\mathcal{E}f(x)|^p\, dx$

$\qquad\qquad \leq C \mathcal{N}_\beta^p(f, F)^p = C \int_F \int_F |f(x) - f(y)|^p \dfrac{d(x, y)^{s - \beta p}}{|B(x, d(x, y))|}\, d\mu(y)\, d\mu(x).$

9.2. PROOF OF THEOREM 9.1

The proof of (a) is divided in two steps. First, we fix $x \in \Omega \setminus F$ and let $B_{i_o} \in \mathcal{F}$ be any (fixed) ball containing x. One has from (9.4)

$$(9.5) \qquad |\mathcal{E}f(x)| \leq \sum_{B_j \in \mathcal{F}'} \phi_j(x) \frac{1}{\mu(\alpha_1 B_j^*)} \int_{\alpha_1 B_j^*} |f(t)| \, d\mu(t)$$

$$\leq \sum_{B_j \in \mathcal{F}'} \phi_j(x) \left(\frac{1}{\mu(\alpha_1 B_j^*)} \int_{\alpha_1 B_j^*} |f(t)|^p \, d\mu(t) \right)^{\frac{1}{p}}$$

$$\leq N \left(\frac{1}{\mu(B_{i_o})} \int_{\alpha_2 B_{i_o}} |f(t)|^p \, d\mu(t) \right)^{\frac{1}{p}}.$$

In the above, we have used (9.3), and the fact that the sum is actually a finite sum of no more than N terms. Next, we obtain from (9.5)

$$(9.6)$$

$$\int_{\Omega \setminus F} |\mathcal{E}f(x)|^p \, dx \leq \sum_{B_i \in \mathcal{F}} \int_{B_i} |\mathcal{E}f(x)|^p \, dx$$

$$\leq \sum_{B_i \in \mathcal{F}} N^p \int_{B_i} \left(\frac{1}{\mu(B_i)} \int_{\alpha_2 B_i} |f(t)|^p \, d\mu(t) \right) dx$$

$$\leq N^p \sum_{B_i \in \mathcal{F}} \frac{|B_i|}{\mu(B_i)} \int_{\alpha_2 B_i} |f(t)|^p \, d\mu(t)$$

$$(\text{by } (1.8)) \leq M \, N^p \sum_{B_i \in \mathcal{F}} r(B_i)^s \int_{\alpha_2 B_i} |f(t)|^p \, d\mu(t)$$

$$\leq C \, \min(dist(F, \partial\Omega), R_o)^s \int_{\alpha_2 B_i} \sum_{B_i \in \mathcal{F}'} \chi_{\alpha_2 B_i}(t) |f(t)|^p \, d\mu(t)$$

$$\leq C \, \min(dist(F, \partial\Omega), R_o)^s \int_F |f(t)|^p \, d\mu(t).$$

We have thus proved (a), with C depending only on M, $dist(F, \partial\Omega)$, and on the characteristic local parameters of U. We now turn to the proof of (b). Let $x \in \Omega \setminus F$. Recall that $\sum_i \phi_i(x) \equiv 1$, and therefore $X_k(\sum_i \phi_i(x)) \equiv X_k(1) \equiv 0$. Using these facts, for any point $y \in \Omega \setminus F$ such that $\mathcal{E}f(y) \neq 0$, we estimate

$$X_k \mathcal{E}f(x)$$

$$= X_k \left(\sum_j \phi_j(x) \frac{1}{\mu(\alpha_1 B_j^*)} \int_{\alpha_1 B_j^*} (f(t) - \mathcal{E}f(y)) \, d\mu(t) \right)$$

$$= \sum_j X_k \phi_j(x) \frac{1}{\mu(\alpha_1 B_j^*)} \int_{\alpha_1 B_j^*} \left(f(t) - \sum_l \phi_l(y) \frac{1}{\mu(\alpha_1 B_l^*)} \int_{\alpha_1 B_l^*} f(\tau) \, d\mu(\tau) \right) d\mu(t)$$

$$= \sum_j X_k \phi_j(x) \left(\sum_l \phi_l(y) \right) \frac{1}{\mu(\alpha_1 B_j^*)} \int_{\alpha_1 B_j^*} f(t) \, d\mu(t)$$

$$- \sum_j \sum_l X_k \phi_j(x) \phi_l(y) \frac{1}{\mu(\alpha_1 B_l^*)} \int_{\alpha_1 B_l^*} f(\tau) \, d\mu(\tau)$$

$$= \sum_j \sum_l X_k \phi_j(x) \phi_l(y) \frac{1}{\mu(\alpha_1 B_j^*)} \int_{\alpha_1 B_j^*} f(t) \, d\mu(t)$$

$$- \sum_j \sum_l X_k \phi_j(x) \phi_l(y) \frac{1}{\mu(\alpha_1 B_l^*)} \int_{\alpha_1 B_l^*} f(\tau) \, d\mu(\tau)$$

$$= \sum_j \sum_l X_k \phi_j(x) \phi_l(y) \left\{ \frac{1}{\mu(\alpha_1 B_j^*)} \int_{\alpha_1 B_j^*} f(t) \, d\mu(t) - \frac{1}{\mu(\alpha_1 B_l^*)} \int_{\alpha_1 B_l^*} f(\tau) \, d\mu(\tau) \right\}$$

$$= \sum_j \sum_l X_k \phi_j(x) \phi_l(y) \mu(\alpha_1 B_j^*)^{-1} \mu(\alpha_1 B_l^*)^{-1}$$

$$\left\{ \mu(\alpha_1 B_l^*) \int_{\alpha_1 B_j^*} f(t) \, d\mu(t) - \mu(\alpha_1 B_j^*) \int_{\alpha_1 B_l^*} f(\tau) \, d\mu(\tau) \right\}$$

$$= \sum_j \sum_l X_k \phi_j(x) \phi_l(y) \mu(\alpha_1 B_j^*)^{-1} \mu(\alpha_1 B_l^*)^{-1}$$

$$\int_{\alpha_1 B_j^*} \int_{\alpha_1 B_l^*} \{f(t) - f(\tau)\} \, d\mu(t) \, d\mu(\tau).$$

If we take $y = x$ in the above, we obtain

(9.7) $\quad |X_k \mathcal{E} f(x)| \leq \sum_j \sum_l |X_k \phi_j(x)| \, \phi_l(x) \cdot$

$$\left(\frac{1}{\mu(\alpha_1 B_j^*)} \frac{1}{\mu(\alpha_1 B_l^*)} \int_{\alpha_1 B_j^*} \int_{\alpha_1 B_l^*} |f(t) - f(\tau)|^p \, d\mu(t) \, d\mu(\tau) \right)^{\frac{1}{p}}.$$

For any fixed $x \in \Omega \setminus F$, and any fixed $B_{i_o} \in \mathcal{F}$ containing x, the sum in (9.7) is over all balls B_j, B_l such that $B_j \cap B_{i_o} \neq \varnothing$, $B_l \cap B_{i_o} \neq \varnothing$, and it is actually finite, with the number of terms independent of x. Hence, from (9.7), and (iii) in Theorem 9.2, we obtain

(9.8)

$|X_k \mathcal{E} f(x)|$

$$\leq \sum_{j=1}^N \sum_{l=1}^N \frac{1}{r(B_j)} \left(\frac{1}{\mu(B_{i_o})^2} \int_{\alpha_2 B_{i_o}} \int_{\alpha_2 B_{i_o}} |f(t) - f(\tau)|^p \, d\mu(t) \, d\mu(\tau) \right)^{\frac{1}{p}}$$

(by Theorem 8.1) $\leq C \dfrac{1}{r(B_{i_o})} \left(\dfrac{1}{\mu(B_{i_o})^2} \displaystyle\int_{\alpha_2 B_{i_o}} \int_{\alpha_2 B_{i_o}} |f(t) - f(\tau)|^p \, d\mu(t) \, d\mu(\tau) \right)^{\frac{1}{p}}.$

Using (9.8) we conclude

(9.9)
$$\int_{\Omega\setminus F} |X_k \mathcal{E} f(x)|^p \, dx$$
$$\leq \sum_{B_i \in \mathcal{F}} \int_{B_i} |X_k \mathcal{E} f(x)|^p \, dx$$
(by (9.8))
$$\leq C \sum_{B_i \in \mathcal{F}} \int_{B_i} \left(\frac{1}{r(B_i)^p} \frac{1}{\mu(B_i)^2} \int_{\alpha_2 B_i} \int_{\alpha_2 B_i} |f(t) - f(\tau)|^p \, d\mu(t) \, d\mu(\tau) \right) dx$$
$$= C \sum_{B_i \in \mathcal{F}} |B_i| \frac{1}{r(B_i)^p} \frac{1}{\mu(B_i)^2} \int_{\alpha_2 B_i} \int_{\alpha_2 B_i} |f(t) - f(\tau)|^p \, d\mu(t) \, d\mu(\tau)$$
(by (1.8)) $\leq C M^2 \sum_{B_i \in \mathcal{F}} \frac{r(B_i)^{2s-p}}{|B_i|} \int_{\alpha_2 B_i} \int_{\alpha_2 B_i} |f(t) - f(\tau)|^p \, d\mu(t) \, d\mu(\tau)$

(by Proposition 1.15)
$$\leq C \sum_{B_i \in \mathcal{F}} \int_{\alpha_2 B_i} \int_{\alpha_2 B_i} |f(t) - f(\tau)|^p \frac{d(t,\tau)^{2s-p}}{|B(t, d(\tau,t))|} \, d\mu(t) \, d\mu(\tau)$$

(by Theorem 8.1) $\leq C M^2 \int_F \int_F |f(t) - f(\tau)|^p \frac{d(t,\tau)^{2s-p}}{|B(t, d(\tau,t))|} \, d\mu(t) \, d\mu(\tau).$

In the second to the last inequality above, we have used the doubling condition (1.22) and the fact that $t, \tau \in B_i$. Also, we have applied Proposition 1.15 with $\alpha = 2s - p$. This is possible provided that $\alpha \leq n$. When $1 \leq p \leq n$, one has equivalently $p \leq \frac{n+p}{2}$, and therefore the condition $\alpha \leq n$ is automatically guaranteed by the assumption $0 < s < p$. This is no longer the case in the range $p > n$, but now the hypothesis $s \leq \frac{n+p}{2}$ implies $\alpha \leq n$. At this point, summing over $k = 1, .., m$ in (9.9) we obtain (b). This concludes the proof of thorem.

REMARK 9.5.

(i) We note that, interestingly, the Poincaré inequality Theorem 7.1 is not needed to establish Theorem 9.1. This is due to the fact that one only needs to control the Sobolev norm from above by the Besov norm.

(ii) As we will see subsequently, in the important situation when F is the boundary of a smooth domain, the parameter s in (1.8) will be $s = 1$, thus the limitation $0 < s < p$ is always satisfied if we confine attention to the range $p > 1$. Moreover, since $n \geq 3$ one also has $\frac{n+p}{2} > 2 > 1 = s$.

(iii) As the end of the proof of Theorem 9.1 shows, the limitation $s \leq \frac{n+p}{2}$ when $p > n$ in its statement comes from the limitation $\alpha \leq n$ in Proposition 1.15. An important situation in which such constraint can be considerably improved is that of a Carnot group $\boldsymbol{G}$ with homogeneous dimension Q. In such case, $|B(x,r)| = \omega \, r^Q$, where $\omega = \omega(\boldsymbol{G}) > 0$ is independent of $x \in \boldsymbol{G}$, and therefore Proposition 1.15 holds trivially with $\alpha \leq Q$. As a consequence, the range of s is wider.

CHAPTER 10

Traces on the boundary of (ϵ, δ) domains

In Chapter 8 we proved that a Sobolev function on a domain Ω possesses a trace on the support F of an upper Ahlfors measure μ, in the situation when F is contained in the interior of Ω. We are presently interested in the question of traces in the important situation when $F = \partial \Omega$. This problem is more delicate than the interior one since it is of a global nature. To explain this point, let us recall that while it is always possible to locally approximate a Sobolev function in the space $\mathcal{L}^{1,p}(\Omega)$ by smooth functions (see [**Fr44**], [**GN96**] and [**FSS96**]), the same is not true up to the boundary of the domain Ω, unless the latter possesses certain geometrical properties. This aspect was investigated in [**GN98**], where, among other things, it was proved that when Ω is a (ϵ, δ)–domain, then the class $C^\infty(\overline{\Omega})$ is dense in $\mathcal{L}^{1,p}(\Omega)$. The purpose of this chapter is to prove that, when μ is an upper Ahlfors measure, the (ϵ, δ) condition also suffices to guarantee that functions in $\mathcal{L}^{1,p}(\Omega, dx)$ have a trace in the Besov space $B^p_{1-s/p}(\partial \Omega, d\mu)$, see Theorem 10.6. Furthermore, if μ is a Ahlfors measure, then such Besov space actually characterizes the trace space on $\partial \Omega$ of Sobolev functions in Ω, see Theorem 10.9. We need to develop some preparatory results.

DEFINITION 10.1. An open set $\Omega \subset \mathbb{R}^n$ is said to have *interior positive density* at $x_o \in \partial \Omega$, if one has

$$D_+(\Omega, x_o) = \limsup_{r \to 0} \frac{|\Omega \cap B(x_o, r)|}{|B(x_o, r)|} > 0.$$

PROPOSITION 10.2. *If Ω has interior positive density at every $x_o \in \partial \Omega$, then $|\partial \Omega| = 0$.*

Proof. Consider the characteristic function χ_Ω. Since $\chi_\Omega \in L^1_{loc}(\mathbb{R}^n)$, by Lebesgue differentiation theorem for spaces of homogeneous type [**St93**], we know that dx-a.e. point of $\mathbb{R}^n$ is a Lebesgue point for χ_Ω. Therefore, the set

$$\mathcal{S} = \left\{ x \in \mathbb{R}^n \mid \frac{|\Omega \cap B(x, r)|}{|B(x, r)|} = \frac{1}{|B(x, r)|} \int_{B(x, r)} \chi_\Omega(y) dy \not\to \chi_\Omega(x), \text{as } r \to 0 \right\}$$

has zero Lebesgue measure. Since the condition $D_+(\Omega, x_o) > 0$ for every $x_o \in \partial \Omega$, implies that $\partial \Omega \subset \mathcal{S}$, the conclusion follows. $\square$

A stronger notion than that of interior positive density is contained in the following definition.

DEFINITION 10.3. An open set $\Omega \subset \mathbb{R}^n$ is called of *type-A* if for $A > 0$, there exists $R_o > 0$ such that for every $x_o \in \partial\Omega$, and $0 < r < R_o$,
$$|\Omega \cap B(x_o, r)| \geq A\,|B(x_o, r)|.$$

Clearly, if Ω is of type-A, then for every $x_o \in \partial\Omega$ one has $D_+(\Omega, x_o) \geq A$. We next recall Definition 9.3, and also (9.2). The following proposition has important potential theoretic consequences.

PROPOSITION 10.4. *If Ω is an (ϵ, δ)-domain, then Ω is type-A with $A = A(\epsilon, C_1)$. In particular, one has $|\partial\Omega| = 0$.*

Proof. Fix $x_o \in \partial\Omega$, and let $y \in \Omega$ be such that $d(x_o, y) < \frac{16}{17}\delta$. Consider $B(x_o, r)$ with $r \leq \frac{1}{2}d(x_o, y) < \frac{8}{17}\delta$, and choose $x \in \Omega$ such that $d(x, x_o) < \frac{r}{8}$. By the triangle inequality, we have $d(x, y) \leq d(x, x_o) + d(x_o, y) \leq \delta$. Definition 9.3 guarantees the existence of a continuous curve $\gamma : [0, T] \to \Omega$, such that $\gamma(0) = x$, $\gamma(T) = y$, and satisfying (9.1). Consider the continuous function $g : [0, T] \to \mathbb{R}$ given by $g(t) = d(\gamma(t), \gamma(0))$. Since $g(0) = 0$ and

(10.1) $$g(T) = d(x, y) \geq d(x_o, y) - d(x_o, x) > 2r - \frac{r}{8} > \frac{r}{8},$$

by the intermediate value theorem, there exists $z \in \{\gamma\}$ such that $d(x, z) = \frac{r}{8}$. One has from (9.1)

$$d(z, \partial\Omega) \geq \epsilon \min\{d(z, x), d(z, y)\} \geq \epsilon \min\{d(z, x), d(x, y) - d(x, z)\}$$
$$> \epsilon \min\{\frac{r}{8}, 2r - \frac{r}{8} - \frac{r}{8}\} = \epsilon\frac{r}{8}.$$

This implies $B(z, \epsilon\frac{r}{8}) \subset \Omega$. Furthermore, since $d(z, x_o) \leq d(z, x) + d(x, x_o) \leq \frac{r}{4}$ and hence, $z \in B(x_o, r/4)$. We infer

$$B(z, \epsilon\frac{r}{8}) \subset B(x_o, r) \cap \Omega, \qquad B(x_o, r) \subset B(z, \frac{5}{4}r).$$

Using Theorem 1.12 we conclude

$$\frac{|\Omega \cap B(x_o, r)|}{|B(x_o, r)|} \geq \frac{|B(z, \frac{\epsilon r}{8})|}{|B(x_o, r)|} \geq A(\epsilon, C_1)\frac{|B(z, \frac{5}{4}r)|}{|B(x_o, r)|} \geq A(\epsilon, C_1) > 0.$$

This completes the proof. $\square$

REMARK 10.5. Due to the nature of the (ϵ, δ) condition, Proposition 10.4 represents a useful tool when one wants to show that a given domain is not (ϵ, δ). We will see an instance of this in Proposition 10.8 below.

THEOREM 10.6 (**Trace theorem on the boundary**). *Let $U \subset \mathbb{R}^n$ be a bounded set with characteristic local parameters C_1, R_o, and let $p > 1$. There is $\sigma = \sigma(X, U) > 0$ such that, if $\Omega \subset U$ is a bounded (ϵ, δ)-domain with $rad(\Omega) > 0$, $diam(\Omega) < \frac{R_o}{2\sigma}$, $dist(\Omega, \partial U) > R_o$, and μ is an upper s-Ahlfors measure for some $0 < s < p$, having $supp\,\mu \subseteq \partial\Omega$, then for every $0 < \beta \leq 1 - s/p$ there exist a linear operator*
$$Tr : \mathcal{L}^{1,p}(\Omega, dx) \to B_\beta^p(\partial\Omega, d\mu),$$
and a constant $C = C(U, p, s, M, \beta, \epsilon, \delta, rad(\Omega)) > 0$, such that

(10.2) $$\|Tr\,f\|_{B_\beta^p(\partial\Omega, d\mu)} \leq C\,\|f\|_{\mathcal{L}^{1,p}(\Omega, dx)}.$$

Furthermore, if $f \in C^\infty(\overline{\Omega}) \cap \mathcal{L}^{1,p}(\Omega, dx)$, then $\mathcal{T}r\, f = f$ on $\partial\Omega$.

Proof. Let $\sigma = \sigma(X, U) > 0$ be the parameter whose existence is asserted by Theorem 8.6. With $f \in \mathcal{L}^{1,p}(\Omega, dx)$ given, we fix $x_o \in \Omega$. In what follows, we continue to indicate with f the precise representation $f^\star$, see Definition 7.4. If $R = diam(\Omega)$, one clearly has $\Omega \subset B_o = B(x_o, R)$. The assumption $diam(\Omega) < \frac{R_o}{2\sigma}$ implies $0 < R < R_o/\sigma$. Since by hypothesis $dist(\Omega, \partial U) > R_o$, we conclude $\sigma B_o \subset U$. Let $\mathcal{S} : \mathcal{L}^{1,p}(\Omega, dx) \to \mathcal{L}^{1,p}(\sigma B_o, dx)$ be a Sobolev extension operator given by Theorem 9.4. We next consider the set

$$\mathcal{G} = \left\{ x \in \partial\Omega \,\Big|\, \lim_{r \to 0} \frac{1}{|B(x,r) \cap \Omega|} \int_{B(x,r) \cap \Omega} f(y)\, dy \text{ exists} \right\}.$$

Our first objective is to prove that

(10.3) $$\mu\,(\partial\Omega \setminus \mathcal{G}) = 0.$$

To this end, we show

$$\mathcal{E} = \left\{ x \in \partial\Omega \,\Big|\, \lim_{r \to 0} \frac{1}{|B(x,r)|} \int_{B(x,r)} |\mathcal{S}f(y) - \mathcal{S}f(x)|\, dy = 0 \right\} \subset \mathcal{G}.$$

In fact, suppose for a moment we have proved the latter inclusion. Theorem 7.8 allows to conclude $\mu(\partial\Omega \setminus \mathcal{E}) = 0$, and therefore (10.3) would follow from $\partial\Omega \setminus \mathcal{G} \subset \partial\Omega \setminus \mathcal{E}$. Let thus $x \in \mathcal{E}$. Proposition 10.4 gives

$$\lim_{r \to 0} \frac{1}{|B(x,r) \cap \Omega|} \int_{B(x,r) \cap \Omega} |f(y) - \mathcal{S}f(x)|\, dy$$

$$= \lim_{r \to 0} \frac{1}{|B(x,r) \cap \Omega|} \int_{B(x,r) \cap \Omega} |\mathcal{S}f(y) - \mathcal{S}f(x)|\, dy$$

$$\leq \lim_{r \to 0} \frac{|B(x,r)|}{|B(x,r) \cap \Omega|} \frac{1}{|B(x,r)|} \int_{B(x,r)} |\mathcal{S}f(y) - \mathcal{S}f(x)|\, dy$$

$$= 0.$$

This shows that

(10.4) $$\lim_{r \to 0} \frac{1}{|B(x,r) \cap \Omega|} \int_{B(x,r) \cap \Omega} f(y)\, dy = \mathcal{S}f(x)$$

(recall that $Sf = Sf^\star$, and therefore $Sf(x)$ exists for every x). We have thus proved that $x \in \mathcal{G}$, and therefore (10.3) is valid. Next, we define for $x \in \partial\Omega$

$$\mathcal{T}r\, f(x) \stackrel{def}{=} \lim_{r \to 0} \frac{1}{|B(x,r) \cap \Omega|} \int_{B(x,r) \cap \Omega} f(y)\, dy.$$

According to (10.3), (10.4), the function $\mathcal{T}r\, f$ is defined $\mu - a.e.$ on $\partial\Omega$. Moreover, (10.4) implies that the definition of $\mathcal{T}r$ is independent of the choice of the operator $\mathcal{S}$ (although $\mathcal{S}f$ was used to show that the trace operator $\mathcal{T}r$ is well defined). Finally, if $f \in C^\infty(\overline{\Omega}) \cap \mathcal{L}^{1,p}(\Omega, dx)$, then $\mathcal{T}r\, f(x) = f(x)$, for all $x \in \partial\Omega$. Our final task is to prove that $Tr f \in B^p_\beta(\partial\Omega, d\mu)$. Using Proposition 10.4 we obtain for $x \in \mathcal{G}$

$$\left| \frac{1}{|B(x,r) \cap \Omega|} \int_{B(x,r) \cap \Omega} f(y) \, dy \right|$$
$$\leq \frac{|B(x,r)|}{|B(x,r) \cap \Omega|} \frac{1}{|B(x,r)|} \int_{B(x,r)} |\mathcal{S}f(y)| \, dy$$
$$\leq \frac{1}{A} \frac{1}{|B(x,r)|} \int_{B(x,r)} |\mathcal{S}f(y)| \, dy.$$

Passing to the limit as $r \to 0$, and using again Theorem 7.8, we conclude

(10.5) $$Tr f(x) \leq \frac{1}{A} \mathcal{S}f(x), \quad \text{for} \quad \mu - \text{a.e. } x \in \partial\Omega.$$

Using in succession (10.5), Theorem 8.6, and Theorem 9.4, we find

$$\|Tr f\|_{B^p_\beta(\partial\Omega, d\mu)} \leq \frac{1}{A} \|\mathcal{S}f\|_{B^p_\beta(\partial\Omega, d\mu)} \leq C \|\mathcal{S}f\|_{\mathcal{L}^{1,p}(\sigma B_o, dx)} \leq C \|f\|_{\mathcal{L}^{1,p}(\Omega, dx)}.$$

This completes the proof. $\square$

The following theorem has a special interest in the applications. Its proof is a direct consequence of Theorems 5.6, and 10.6.

THEOREM 10.7. *Let $U \subset \mathbb{R}^n$ be a bounded set with characteristic local parameters C_1, R_o, and let $p > 1$. There is $\sigma = \sigma(X, U) > 0$ such that, if $\Omega \subset U$ is a $C^{1,1}$, (ϵ, δ)-domain with $rad(\Omega) > 0, diam(\Omega) < \frac{R_o}{2\sigma}, dist(\Omega, \partial U) > R_o$, and μ is the perimeter measure on $\partial\Omega$, then if Ω is of type ≤ 2 there exist a linear operator*

$$Tr \ : \ \mathcal{L}^{1,p}(\Omega, dx) \ \to \ B^p_{1-\frac{1}{p}}(\partial\Omega, d\mu) \ ,$$

and a constant $C = C(U, p, s, M, \epsilon, \delta, rad(\Omega)) > 0$, such that

(10.6) $$\|Tr f\|_{B^p_{1-\frac{1}{p}}(\partial\Omega, d\mu)} \leq C \|f\|_{\mathcal{L}^{1,p}(\Omega, dx)} \ .$$

Furthermore, if $f \in C^\infty(\overline{\Omega}) \cap \mathcal{L}^{1,p}(\Omega, dx)$, then $Tr f = f$ on $\partial\Omega$.

10.1. The (ϵ, δ) condition is optimal for the existence of traces

In the sequel we prove that the (ϵ, δ) assumption in Theorem 10.6 cannot be weakened. What we mean by this is that there exist $C^{1,\alpha}$ domains in Carnot groups of step 2 which fail to be (ϵ, δ), and for which it is impossible to define the trace on the boundary of Sobolev functions. To prove this negative phenomenon we consider the domain Ω introduced in (6.39). In the sequel we continue to use the notations of the sub-chapter 6.5.

PROPOSITION 10.8. *The domain $\Omega \subset \mathbb{H}^1$ defined in (6.39) is not a (ϵ, δ)-domain. Moreover, there exist a function $f \in \mathcal{L}^{1,2}(\Omega, dg)$ whose trace does not belong to the Besov space $B^2_{1/2}(\partial\Omega, d\mu)$, where $d\mu$ denotes the perimeter measure defined in Definition 4.7.*

10.1. THE (ϵ, δ) CONDITION IS OPTIMAL FOR THE EXISTENCE OF TRACES

PROOF. To prove that Ω is not (ϵ, δ) it suffices to show, in view of Proposition 10.4, that Ω fails to be of type A. To this purpose it is enough to prove that

$$(10.7) \qquad \lim_{r \to 0} \frac{|\Omega \cap Box(e, r)|}{|Box(e, r)|} = 0 \,.$$

It is now easy to see that

$$|\Omega \cap Box(e, r)| = \pi \int_0^{r^2} t^{2/\beta}\, dt = \pi \frac{\beta}{\beta + 2} r^{2 + \frac{4}{\beta}} \,.$$

On the other hand $|Box(e, r)| = 2\pi r^4$, and therefore

$$\frac{|\Omega \cap Box(e, r)|}{|Box(e, r)|} = \frac{\beta}{2(\beta + 2)} r^{\frac{4}{\beta} - 2} \,.$$

Since $4/\beta - 2 > 0$, the latter equation proves (10.7). We notice that when $\beta = 2$ the domain Ω is of type A. In fact, thanks to the results in [**CG98**], in this case Ω is a NTA domain, and therefore also (ϵ, δ). We next show that Theorem 10.6 fails to be valid for Ω. For simplicity we consider the case $p = 2$. Let $u(g) = t^{-a}$, with $a > 0$ to be chosen. We claim that if

$$(10.8) \qquad \frac{1}{\beta} \le a < \frac{2}{\beta} - \frac{1}{2} \,,$$

then $u \in \mathcal{L}^{1,2}(\Omega, dg)$, but the trace of u does not belong to the Besov space $B^2_{1/2}(\partial\Omega, d\mu)$. We note explicitly that when $\beta = 2$ the numbers in left- and right-hand sides of (10.8) coincide, and the interval is empty. In order to prove that $u \in \mathcal{L}^{1,2}(\Omega, dg)$, we compute $Xu = (X_1 u, X_2 u)$ using (3.19),

$$X_1 u(g) = -2\,a\,y\,t^{-a-1} \,, \qquad X_2 u(g) = 2\,a\,x\,t^{-a-1} \,.$$

This gives

$$\int_\Omega |Xu|^2\, dg = \int_0^1 \int_{|z| < t^{1/\beta}} 4a^2 |z|^2\, t^{-2(a+1)}\, dz\, dt \le 4a^2 \int_0^1 t^{4/\beta - 2(a+1)}\, dt \,.$$

It is then clear that $u \in \mathcal{L}^{1,2}(\Omega, dg)$ provided that the second inequality in (10.8) hold. We next want to show that if a satisfies the left-hand side inequality in (10.8), then
$$(10.9)$$
$$\mathcal{N}^2_\beta(u, F, d\mu) = \left\{ \int_F \int_F \left(\frac{|u(g) - u(g')|}{d(g, g')^{1/2}} \right)^2 \frac{d(g, g')}{|B(g, d(g, g'))|} d\mu(g)\, d\mu(g') \right\}^{\frac{1}{2}} = \infty,$$

where $F \subset \partial\Omega$ is as in chapter 6.5, and $\mathcal{N}^2_\beta(u, F, d\mu)$ is the semi-norm of u in the Besov space $B^2_{1/2}(F, d\mu)$. Recalling that $d(g, g') \cong N(g \circ g'^{-1})$, where $N(g)$ is the gauge in (3.21), and the group law is given by (3.20), we see that in order to prove (10.9) it suffices to show

$$(10.10) \quad \mathcal{N}^* \stackrel{def}{=} \int_F \int_F \frac{|u(g) - u(g')|^2}{N(g \circ g'^{-1})^4} |X\phi(g)|\, |X\phi(g')|\, d\sigma(g)\, d\sigma(g') = \infty \,.$$

Keeping in mind (6.42), we have

$$\mathcal{N}^* \ge C(\beta) \int_{|z| < 1} \int_{|\zeta| < 1} \frac{||z|^{-a\beta} - |\zeta|^{-a\beta}|^2}{N(g \circ g'^{-1})^4} |z|^{\beta - 1} |\zeta|^{\beta - 1}\, dz\, d\zeta \,,$$

where
$$N(g \circ g'^{-1})^4 = (|z|^2 + |\zeta|^2 - 2<z,\zeta>)^2 + (|z|^\beta - |\zeta|^\beta + 2(x\eta - \xi y))^2 ,$$
and we have let $g' = (\zeta, \tau) = (\xi, \eta, \tau)$. Switching to polar coordinates we find
$$\mathcal{N}^* \geq C(\beta) \int_0^1 \int_0^{2\pi} \int_0^1 \int_0^{2\pi} \frac{|r^{-\alpha\beta} - \rho^{-\alpha\beta}|^2}{A(r,\theta,\rho,\phi)} r^\beta \rho^\beta \, dr \, d\theta \, d\rho \, d\phi \stackrel{def}{=} \mathcal{M},$$
where
$$A(r,\theta,\rho,\phi) = (r^2 + \rho^2 - 2r\rho\cos(\theta - \phi))^2 + (r^\beta - \rho^\beta + 2r\rho\sin(\theta - \phi))^2 .$$
The change of variable $r = \rho s$ in the integral with respect to r gives
$$\mathcal{M} = \int_0^1 \int_0^{2\pi} \int_0^{2\pi} \int_0^{1/\rho} \frac{\rho^{-2\alpha\beta}|1 - s^{-\alpha\beta}|^2 s^\beta \rho^{2\beta+1}}{A(\rho s, \theta, \rho, \phi)} \, ds \, d\theta \, d\phi \, d\rho$$
$$\geq \int_0^1 \int_0^{2\pi} \int_0^{2\pi} \int_0^1 \frac{\rho^{-2\alpha\beta}|1 - s^{-\alpha\beta}|^2 s^\beta \rho^{2\beta+1}}{A(\rho s, \theta, \rho, \phi)} \, ds \, d\theta \, d\phi \, d\rho$$

We now notice that

$$A(\rho s, \theta, \rho, \phi)$$
$$= \rho^4 \left\{s^2 - 2s\cos(\theta - \phi) + 1\right\}^2 + \rho^{2\beta} \left\{s^\beta - 1 + 2s\rho^{2-\beta}\sin(\theta - \phi)\right\}^2$$
$$= \rho^{2\beta} \left\{\rho^{4-2\beta}\left\{s^2 - 2s\cos(\theta - \phi) + 1\right\}^2 + \left\{s^\beta - 1 + 2s\rho^{2-\beta}\sin(\theta - \phi)\right\}^2\right\}$$
$$\stackrel{def}{=} \rho^{2\beta} B(\rho, \theta, s, \phi) .$$

Thanks to the assumption $1 < \beta < 2$, we readily recognize that
$$|B(\rho, \theta, s, \phi)| \leq C , \qquad 0 \leq \rho, s \leq 1, 0 \leq \theta, \phi \leq 2\pi ,$$
therefore
$$\mathcal{M} \geq C \int_0^1 \frac{d\rho}{\rho^{2\alpha\beta-1}} \int_0^1 \frac{(1 - s^{\alpha\beta})}{s^{2\alpha\beta-\beta}} \, ds .$$
The integral in the right-hand side of the latter inequality is divergent provided that $2\alpha\beta - 1 \geq 1$, i.e., $\alpha \geq 1/\beta$, which is the inequality in the left-hand side of (10.8). This proves (10.10), so the proof is completed. $\square$

10.2. Characterization of the traces on the boundary

In what follows we finally prove the central result of this chapter, namely, the characterization of the trace space for a Sobolev space.

THEOREM 10.9. *Let $U \subset \mathbb{R}^n$ be a bounded set, with characteristic local parameters C_1, R_o, and $\Omega \subset U$ be an (ϵ, δ)-domain, with $rad(\Omega) > 0$, $dist(\Omega, \partial U) > R_o$. Given $p > 1$, let μ be a s-Ahlfors measure, with $0 < s < p$ and $supp \, \mu \subseteq \partial \Omega$. When $p > n$ we assume in addition that $s \leq \frac{n+p}{2}$. There is $\sigma = \sigma(U, X) > 0$ such that, if $diam(\Omega) < \frac{R_o}{2\sigma}$, then there exist two linear operators*
$$\mathcal{T}r : \mathcal{L}^{1,p}(\Omega, dx) \to B^p_{1-s/p}(\partial\Omega, d\mu) , \qquad \mathcal{E} : B^p_{1-s/p}(\partial\Omega, d\mu) \to \mathcal{L}^{1,p}(\Omega, dx) ,$$
such that $\mathcal{T}r \circ \mathcal{E}$ is the identity map from $B^p_{1-s/p}(\partial\Omega, d\mu)$ into itself.

10.2. CHARACTERIZATION OF THE TRACES ON THE BOUNDARY

Proof. We begin by observing that, thanks to Proposition 10.4, we have $|\partial\Omega| = 0$, therefore if $F = supp\ \mu$, we obtain in particular $|F| = 0$. Since μ is, in particular, a lower s-Ahlfors measure, with s within the range of the hypothesis of Theorem 9.1, we can apply this result which guarantees the existence of an extension operator

$$\mathcal{E}\ :\ B^p_{1-s/p}(\partial\Omega, d\mu)\ \longrightarrow\ \mathcal{L}^{1,p}(\tilde{\Omega}, dx)\ ,$$

where $\tilde{\Omega}$ is a bounded open set such that $\overline{\Omega} \subset \tilde{\Omega}$. Clearly, $\mathcal{E}f\mid_{\Omega} \in \mathcal{L}^{1,p}(\Omega, dx)$. The fact that μ is an upper s-Ahlfors measure allows to apply Theorem 10.6 to infer the existence of a trace operator

$$\mathcal{T}r\ :\ \mathcal{L}^{1,p}(\Omega, dx)\ \longrightarrow\ B^p_{1-s/p}(\partial\Omega, d\mu)\ .$$

This completes the proof. □

Theorem 10.9 gives a complete characterization of the traces of Sobolev functions for the general class of (ϵ, δ) domains. Thanks to such result it is now possible to say that *the space $B^p_{1-s/p}(\partial\Omega, d\mu)$ is the trace space of $\mathcal{L}^{1,p}(\Omega, dx)$ on the boundary of Ω*. We will express this property as follows

$$\mathcal{L}^{1,p}(\Omega, dx)|_{\partial\Omega}\ =\ B^p_{1-s/p}(\partial\Omega, d\mu).$$

We close this chapter with a different trace theorem which follows directly from the cited Theorem 1.1, established in [**DGN98**], and from the extension Theorem 9.4 from [**GN98**].

THEOREM 10.10. *Let $U \subset \mathbb{R}^n$ be a bounded set, with characteristic local parameters C_1, R_o. Consider an (ϵ, δ)-domain $\Omega \subset U$, with $rad(\Omega) > 0$, and an upper s-Ahlfors measure μ for some $0 < s < p$, having $supp\ \mu \subseteq \partial\Omega$, with $\mu(\partial\Omega) > 0$. There is $\sigma = \sigma(U, X) > 0$ such that, if $diam(\Omega) < \frac{R_o}{2\sigma}$, there exists a restriction operator*

$$\mathcal{T}r\ :\ \mathcal{L}^{1,p}(\Omega, dx) \to L^q(\partial\Omega, d\mu) \qquad \text{where } q\ =\ p\frac{Q-s}{Q-p} > p$$

such that $\mathcal{T}rf = f$ for $f \in C^\infty(\overline{\Omega}) \cap \mathcal{L}^{1,p}(\Omega, dx)$ and

$$\|\mathcal{T}rf\|_{L^q(\partial\Omega, d\mu)}\ \leq\ C\ \|f\|_{\mathcal{L}^{1,p}(\Omega, dx)}\ .$$

This result is of course different from the much deeper Theorem 10.6. A comparison, and some comments about these two theorems, are found in Chapter 11.

CHAPTER 11

The embedding of $B^p_\beta(\Omega, d\mu)$ into $L^q(\Omega, d\mu)$

Theorem 1.1 in the introduction claims that a function in the sub-elliptic Sobolev space $\mathcal{L}^{1,p}(\sigma B_o, dx)$ possesses a trace on the support of the measure μ. Furthermore, such trace belongs to a Lebesgue space $L^q(B_o, d\mu)$, with an optimal gain in the exponent of integrability. On the other hand, in Theorem 8.6 we have proved that $\mathcal{L}^{1,p}(\sigma B_o, dx)$ embeds continuously into the Besov space $B^p_\beta(B_o, d\mu)$. The question naturally arises of whether it is possible to close the gap between these two results, by showing that $B^p_\beta(B_o, d\mu) \subset L^q(B_o, d\mu)$. The purpose of this chapter is to prove that this is in fact possible, provided that μ is a lower s-Ahlfors measure. The following theorem is our main result in this direction.

THEOREM 11.1 (**Embedding a Besov space into a Lebesgue space**). *Given a bounded set $U \subset \mathbb{R}^n$ having characteristic local parameters C_1, R_o, and local homogeneous dimension Q, let $\Omega \subset \overline{\Omega} \subset U$ be an open set with diam $\Omega < R_o/2$. Let $p \geq 1$, $0 < \beta < 1$. Suppose μ is a lower s-Ahlfors measure with*

(11.1) $$0 < s \leq n + \beta p, \qquad s < Q - \beta p,$$

and such that supp $\mu = F \subset \Omega$. There exists a continuous embedding

$$B^p_\beta(F, d\mu) \subset L^q(\Omega, d\mu), \qquad \text{where} \quad q = p\,\frac{Q-s}{Q-s-\beta p},$$

and, in fact, for $f \in B^p_\beta(F, d\mu)$ one has

(11.2)

$$\|f\|_{L^q(\Omega,d\mu)} \leq C \left\{ \left(1 + \frac{diam(\Omega)^\beta}{\mu(F)^{\beta/(Q-s)}} \right) \mathcal{N}^p_\beta(f, F, d\mu) + \frac{1}{\mu(F)^{\beta/(Q-s)}}\, \|f\|_{L^p(\Omega,d\mu)} \right\},$$

where $C = C(\Omega, C_1, R_o, p, \beta, s, M) > 0$. Furthermore,

(11.3)

$$\left(\int_\Omega |f(x) - f_{\Omega,\mu}|^q\, d\mu(x) \right)^{\frac{1}{q}} \leq C \left(\int_F \int_F |f(x) - f(y)|^p \frac{d(x,y)^{s-\beta p}}{|B(x, d(x,y))|}\, d\mu(y)\, d\mu(x) \right)^{\frac{1}{p}},$$

where $f_{\Omega,\mu}$ denotes the average $\frac{1}{\mu(\Omega)} \int_\Omega f\, d\mu$.

REMARK 11.2. If in the statement of Theorem 11.1 we take $1 \leq p < Q$, and $\beta = 1 - s/p$ for some $0 < s < p$, then condition (11.1) translates into $s \leq \frac{n+p}{2}$ (the reader should note that such inequality is automatically guaranteed by the condition $s < p$ when $p \leq n$. Thereby, it imposes an additional restriction on s only when $n < p < Q$.) If we assume, in addition to (1.8), that μ satisfy (1.7), then combining Theorem 8.6 with Theorem 11.1, we obtain a stronger version of the cited Theorem 1.1, established in our previous work [**DGN98**]. However, while in Theorem 11.1 we have assumed (1.8), such hypothesis is not needed in Theorem 1.1.

Proof. We divide the proof into several steps. Given $f \in B^p_\beta(\Omega, d\mu)$, for $0 < r < diam(\Omega) < R_o$ consider a ball $B(x_o, r)$ such that $x \in B(x_o, r)$. Using (1.8) and Theorem 1.12, we can estimate

$$|f(x) - f_{B(x_o,r),\mu}| \leq \left\{ \frac{1}{\mu(B(x_o,r))} \int_{B(x_o,r)} |f(x) - f(y)|^p \, d\mu(y) \right\}^{\frac{1}{p}}$$

$$\leq M^{1/p} \left\{ \frac{r^s}{|B(x_o,r)|} \int_{B(x_o,r)} |f(x) - f(y)|^p \, d\mu(y) \right\}^{\frac{1}{p}}$$

$$\leq \left(\frac{C_1 M}{2^s} \right)^{1/p} \left\{ \frac{(2r)^s}{|B(x_o, 2r)|} \int_{B(x_o,r)} |f(x) - f(y)|^p \, d\mu(y) \right\}^{\frac{1}{p}}$$

If we use Proposition 1.15 with $\alpha = s - \beta p \leq n$, we obtain
(11.4)
$$\frac{|f(x) - f_{B(x_o,r),\mu}|}{r^\beta} \leq (CM)^{1/p} \left\{ \int_\Omega |f(x) - f(y)|^p \frac{d(x,y)^{s-\beta p}}{|B(x, d(x,y))|} \, d\mu(y) \right\}^{\frac{1}{p}},$$

where $C = C(C_1, p, \beta, s) > 0$. The estimate (11.4) suggests to consider the following truncated maximal function

$$G^\beta f(x) = \sup \left\{ \frac{|f(x) - f_{B(x_o,r),\mu}|}{r^\beta} \mid x \in B(x_o, r), \, 0 < r < R_o \right\}.$$

From (11.4) it is clear that

(11.5) $\qquad G^\beta f(x) \leq (CM)^{1/p} \left\{ \int_\Omega |f(x) - f(y)|^p \frac{d(x,y)^{s-\beta p}}{|B(x, d(x,y))|} \, d\mu(y) \right\}^{\frac{1}{p}},$

and therefore

(11.6)
$$\int_\Omega |G^\beta f(x)|^p \, d\mu(x) \leq C M \int_\Omega \int_\Omega |f(x) - f(y)|^p \frac{d(x,y)^{s-\beta p}}{|B(x, d(x,y))|} \, d\mu(y) \, d\mu(x)$$
$$= C M \, \mathcal{N}^p_\beta(f, F, d\mu)^p,$$

which proves, in particular, $G^\beta f \in L^p(\Omega, d\mu)$. Consider next $x, y \in \Omega$, and fix a ball $B(x_o, r)$ for which $x, y \in B(x_o, r)$, with $r \leq 2d(x, y)$. One has

(11.7) $\qquad |f(x) - f(y)| \leq |f(x) - f_{B(x_o,r),\mu}| + |f(y) - f_{B(x_o,r),\mu}|$
$\qquad \qquad \leq 2^\beta \, d(x,y)^\beta \left\{ G^\beta f(x) + G^\beta f(y) \right\}.$

We let $g = G^\beta f$. We can assume $g > 0$, otherwise f would be constant and hence $f \in L^q(\Omega, d\mu)$ trivially. If we let $E_k = \{x \in \Omega \mid g(x) \leq 2^k\}$, then we obtain

$$
(11.8) \quad \int_\Omega g(x)^p \, d\mu(x) = p \int_0^\infty t^{p-1} \mu(\{x \in \Omega \mid g(x) > t\}) \, dt
$$

$$
\geq \frac{p}{2^p} \sum_{k \in \mathbb{Z}} 2^{kp} \mu(\{x \in \Omega \mid g(x) > 2^k\})
$$

$$
\geq \frac{p}{2^p} \sum_{k \in \mathbb{Z}} 2^{kp} \, \mu(E_{k+1} \setminus E_k).
$$

Let now $C^* = C(\overline{\Omega}, R_o) > 0$ be the constant in (1.24). If we set

$$
r = \min \left\{ \left(\frac{2MR_o^Q}{C^*}\right)^{\frac{1}{Q-s}} \mu(E_{k-1}^c)^{\frac{1}{Q-s}}, R_o \right\},
$$

then $0 < r \leq R_o$. Since by hypothesis $\text{diam}\,(\Omega) < R_o/2$, it is clear that if $r = R_o$, then for every $x \in E_k$ one has trivially

$$
(11.9) \quad B(x, r) \cap E_{k-1} \neq \emptyset.
$$

If, on the other hand, $r = \left(\frac{2MR_o^Q}{C^*}\right)^{\frac{1}{Q-s}} \mu(E_{k-1}^c)^{\frac{1}{Q-s}}$, then for $x \in E_k$ we can apply (1.8), and (1.24), obtaining
(11.10)

$$
\mu(B(x,r)) \geq M^{-1} \frac{|B(x,r)|}{r^s} \geq M^{-1} \frac{C^*}{R_o^Q} r^{Q-s} = 2\,\mu(E_{k-1}^c) > \mu(E_{k-1}^c).
$$

This implies that (11.9) holds in this case as well. Otherwise, we would have from (11.10)

$$
\mu(E_k^c) + \mu(E_k \setminus E_{k-1}) = \mu(E_{k-1}^c) < \mu(B(x,r))
$$
$$
= \mu(B(x,r) \setminus E_k) + \mu(B(x,r) \cap (E_k \setminus E_{k-1}))
$$
$$
\leq \mu(E_k^c) + \mu(E_k \setminus E_{k-1}),
$$

a contradiction. Let then $\bar{x} \in B(x,r) \cap E_{k-1}$, and define

$$
a_k = \sup_{E_k} |f|.
$$

From (11.7) and the definition of r we obtain for $x \in E_k$

$$
|f(x)| \leq |f(x) - f(\bar{x})| + |f(\bar{x})| \leq C\, d(x, \bar{x})^\beta (g(x) + g(\bar{x})) + \sup_{E_{k-1}} |f|
$$

$$
\leq C\, \mu(E_{k-1}^c)^{\frac{\beta}{Q-s}} (2^k + 2^{k-1}) + \sup_{E_{k-1}} |f|
$$

$$
\leq C\, 2^{k-1} \mu(E_{k-1}^c)^{\frac{\beta}{Q-s}} + a_{k-1}.
$$

Chebychev inequality gives

$$
(11.11) \quad \mu(E_{k-1}^c) = \mu(\{x \in \Omega \mid g(x) > 2^{k-1}\}) \leq \frac{2^p}{2^{kp}} \int_\Omega |g(x)|^p \, d\mu(x),
$$

and using this information in the latter inequality we find

$$
(11.12) \quad a_k \leq C\, 2^{k(1 - \frac{\beta p}{Q-s})} \|g\|_{L^p(\Omega, d\mu)}^{\frac{\beta p}{Q-s}} + a_{k-1},
$$

where $C = C(C_1, R_o, p, s, \beta)$. Using the fact that $\Omega = \bigcup_{k \in \mathbb{Z}} E_k$, and the monotonicity $E_k \subset E_{k+1}$, one can find $k_o \in \mathbb{Z}$ such that

$$\mu(E_{k_o-1}) \leq \frac{\mu(\Omega)}{2} < \mu(E_{k_o}). \tag{11.13}$$

This gives in particular

$$\frac{\mu(\Omega)}{2} < \mu(E_{k_o-1}^c).$$

This inequality and (11.11) imply

$$2^{k_o} \leq C \frac{1}{\mu(\Omega)^{\frac{1}{p}}} \|g\|_{L^p(\Omega, d\mu)}. \tag{11.14}$$

Consider now

$$b_k = \inf_{E_k} |f| \leq \left(\frac{1}{\mu(E_k)} \int_{E_k} |f(x)|^p \, d\mu(x) \right)^{\frac{1}{p}}. \tag{11.15}$$

Let $x_j \in E_k$ be such that $|f(x_j)| \to b_k$, then (11.7) gives for $x \in E_k$

$$|f(x)| \leq C \, d(x, x_j)^\beta \, (g(x) + g(x_j)) + |f(x_j)| \leq C \, diam(\Omega)^\beta \, 2^{k-1} + |f(x_j)|.$$

Letting $j \to \infty$ we infer

$$|f(x)| \leq C \, diam(\Omega)^\beta \, 2^{k-1} + b_k, \qquad x \in E_k,$$

and therefore, in particular,

$$a_{k_o} = \sup_{E_{k_o}} |f| \leq C \, 2^{k_o-1} \, diam(\Omega)^\beta + b_{k_o} \tag{11.16}$$

$$\text{(by (11.15))} \quad \leq C 2^{k_o-1} diam(\Omega)^\beta + \frac{1}{\mu(E_{k_o})^{\frac{1}{p}}} \|f\|_{L^p(\Omega, d\mu)}$$

$$\text{(by (11.14))} \quad \leq C diam(\Omega)^\beta \frac{1}{\mu(\Omega)^{\frac{1}{p}}} \|g\|_{L^p(\Omega, d\mu)} + \frac{1}{\mu(E_{k_o})^{\frac{1}{p}}} \|f\|_{L^p(\Omega, d\mu)}$$

$$\text{(by (11.13))} \quad \leq \frac{C}{\mu(\Omega)^{\frac{1}{p}}} \left(diam(\Omega)^\beta \|g\|_{L^p(\Omega, d\mu)} + \|f\|_{L^p(\Omega, d\mu)} \right).$$

Since the a_k's are increasing in k, if $k \leq k_o$, then $a_k \leq a_{k_o}$. If instead $k > k_o$, then iterating (11.12), and observing that (11.1) implies $1 - \frac{\beta p}{Q-s} > 0$, we obtain
(11.17)

$$a_k \leq C \, \|g\|_{L^p(\Omega, d\mu)}^{\frac{\beta p}{Q-s}} \sum_{j=-\infty}^{k} 2^{j(1-\frac{\beta p}{Q-s})} + a_{k_o} \leq C \, \|g\|_{L^p(\Omega, d\mu)}^{\frac{\beta p}{Q-s}} 2^{k(1-\frac{\beta p}{Q-s})} + a_{k_o}.$$

We conclude from (11.17) that, if $q = p \frac{Q-s}{Q-s-\beta p}$, then

$$\int_\Omega |f(x)|^q \, d\mu$$
$$\leq \sum_{k \in \mathbb{Z}} a_k^q \mu(E_k \setminus E_{k-1})$$
$$\leq C \sum_{k \in \mathbb{Z}} \left[\|g\|_{L^p(\Omega, d\mu)}^{q\frac{\beta p}{Q-s}} 2^{kq(1-\frac{\beta p}{Q-s})} \mu(E_k \setminus E_{k-1}) + a_{k_o}^q \mu(E_k \setminus E_{k-1}) \right]$$
$$\leq C \, \|g\|_{L^p(\Omega, d\mu)}^{q-p} \sum_{k \in \mathbb{Z}} 2^{kp} \mu(E_k \setminus E_{k-1}) + a_{k_o}^q \mu(\Omega)$$

(by (11.8) and (11.16))
$$\leq C \, \|g\|_{L^p(\Omega, d\mu)}^q + C \, (\|f\|_{L^p(\Omega, d\mu)}^q) \mu(\Omega)^{1-\frac{q}{p}} + \, diam(\Omega)^{\beta q} \|g\|_{L^p(\Omega, d\mu)}^q$$
$$\leq C \, (1 + diam(\Omega)^{\beta q} \mu(\Omega)^{1-\frac{q}{p}}) \|g\|_{L^p(\Omega, d\mu)}^q + C \, \mu(\Omega)^{1-\frac{q}{p}} \|f\|_{L^p(\Omega, d\mu)}^q.$$

The above arguments show that $f \in L^q(\Omega, d\mu)$ with
(11.18)
$$\|f\|_{L^q(\Omega, d\mu)} \leq C \left\{ 1 + \frac{diam(\Omega)^\beta}{\mu(\Omega)^{\beta/(Q-s)}} \right\} \|g\|_{L^p(\Omega, d\mu)} + \tilde{C} \frac{1}{\mu(\Omega)^{\beta/(Q-s)}} \|f\|_{L^p(\Omega, d\mu)}.$$

To complete the proof, we are left with establishing (11.3). To this end, we prove
$$(11.19) \qquad \|f - f_{\Omega, \mu}\|_{L^p(\Omega, d\mu)} \leq C \, diam(\Omega)^\beta \|g\|_{L^p(\Omega, d\mu)}.$$

Assuming (11.19) for the moment. Applying (11.18) to $f - f_{\Omega, \mu}$ and observe that $G_h^\beta f(x) = G_h^\beta (f - c)(x)$ for any constant $c \in \mathbb{R}$ and we have

$$\|f - f_{\Omega, d\mu}\|_{L^q(\Omega, d\mu)}$$
$$\text{(by (11.18))} \leq C(1 + diam(\Omega)^\beta \mu(\Omega)^{-\frac{\beta}{Q-s}}) \|g\|_{L^p(\Omega, d\mu)}$$
$$\qquad + \tilde{C} \mu(\Omega)^{-\frac{\beta}{Q-s}} \|f - f_{\Omega, \mu}\|_{L^p(\Omega, d\mu)}$$
$$\text{(by (11.19))} \leq C(1 + diam(\Omega)^\beta \mu(\Omega)^{-\frac{\beta}{Q-s}}) \|g\|_{L^p(\Omega, d\mu)}$$
$$\qquad + \tilde{C} diam(\Omega)^\beta \mu(\Omega)^{-\frac{\beta}{Q-s}} \|g\|_{L^p(\Omega, d\mu)}$$
$$\leq C'(1 + diam(\Omega)^\beta \mu(\Omega)^{-\frac{\beta}{Q-s}}) \|g\|_{L^p(\Omega, d\mu)}.$$

Now, we turn to the proof of (11.19). For $x \in \Omega$ we have
$$|f(x) - f_{\Omega, \mu}| \leq \frac{1}{\mu(\Omega)} \int_\Omega |f(x) - f(y)| \, d\mu(y)$$
$$\text{(by (11.7))} \leq C \frac{diam(\Omega)^\beta}{\mu(\Omega)} \int_\Omega |g(x) + g(y)| \, d\mu(y)$$
$$= C diam(\Omega)^\beta \left(g(x) + \frac{1}{\mu(\Omega)} \int_\Omega |g(y)| \, d\mu(y) \right)$$
$$\leq C diam(\Omega)^\beta \left[g(x) + \left(\frac{1}{\mu(\Omega)} \int_\Omega |g(y)|^p \, d\mu(y) \right)^{\frac{1}{p}} \right].$$

Thus,

$$\|f - f_{\Omega,\mu}\|_{L^p(\Omega,d\mu)}$$
$$= \left(\int_\Omega |f(x) - f_{\Omega,\mu}|^p \, d\mu(x)\right)^{\frac{1}{p}}$$
$$\leq \left(C^p \, diam(\Omega)^{\beta p} \left[\int_\Omega |g(x)|^p \, d\mu(x) + \left(\frac{1}{\mu(\Omega)} \int_\Omega |g(y)|^p \, d\mu(y)\right) \mu(\Omega)\right]\right)^{\frac{1}{p}}$$
$$= 2C \, diam(\Omega)^\beta \left(\int_\Omega |g(x)|^p \, d\mu(x)\right)^{\frac{1}{p}}.$$

This completes the proof of the theorem. $\square$

We end this chapter by showing that, when the measure μ is an s-Ahlfors measure, then the Besov space $B_\beta^p(F, d\mu)$ is a Banach space. In what follows, Q always denotes the local homogeneous dimension corresponding to a bounded set $U \subset \mathbb{R}^n$.

THEOREM 11.3. *Let $U \subset \mathbb{R}^n$, R_o be as in Theorem 1.12, $1 \leq p < Q$ and μ be a s-Ahlfors measure, with $s \leq \frac{1+p}{2}$. Assume $\operatorname{supp} \mu = F \subset B_o \subset\subset U$ for some $B_o = B(x_o, R)$, with $R < \frac{R_o}{2\sigma}$, where σ is the parameter in Theorem 8.6, $|F| = 0$ and $\mu(F) > 0$. If $0 < \beta \leq 1 - s/p$, then the space $B_\beta^p(F, d\mu)$ is a Banach space.*

Proof. We consider the set
$$\mathcal{Z} = \{f \in \mathcal{L}^{1,p}(B(x_o, \sigma R), dx) \mid \|f^\star\|_{B_\beta^p(F, d\mu)} = 0\}.$$

Note that $\mathcal{Z}$ is well defined by Theorem 8.6, since thanks to the latter one has $\|f^\star\|_{B_\beta^p(F,d\mu)} < \infty$ for $f \in \mathcal{L}^{1,p}(B(x_o, \sigma R), dx)$. We claim that $\mathcal{Z}$ is a closed subspace of the Banach space $\mathcal{L}^{1,p}(B(x_o, \sigma R), dx)$. To see this, we take $f_n \in \mathcal{Z}$ and suppose f_n converges to some f in $\mathcal{L}^{1,p}(B(x_o, \sigma R), dx)$. We wish to show that $f \in \mathcal{Z}$. Now

$$\|f^\star\|_{B_\beta^p(B(x_o,\sigma R), dx)} \leq \|f^\star - f_n^\star\|_{B_\beta^p(B(x_o,\sigma R), d\mu)} + \|f_n^\star\|_{B_\beta^p(B(x_o,\sigma R), d\mu)}$$
$$\text{(by Theorem 8.6)} \leq C\|f^\star - f_n^\star\|_{\mathcal{L}^{1,p}(B(x_o,\sigma R), dx)} \longrightarrow 0 \quad \text{as } n \to \infty.$$

We conclude that $f^\star \in \mathcal{Z}$. Hence the quotient space $\mathcal{L}^{1,p}(B(x_o, \sigma R), dx)/\mathcal{Z}$ is also a Banach space, see e.g., [**Sch71**, Theorem 5.2]. Next, we show that there is a continuous bijection between $B_\beta^p(F, d\mu)$ and $\mathcal{L}^{1,p}(B(x_o, \sigma R), dx)/\mathcal{Z}$. We define $\Phi : B_\beta^p(F, d\mu) \to \mathcal{L}^{1,p}(B(x_o, \sigma R), dx)/\mathcal{Z}$ to be

$$\Phi = \pi \circ \mathcal{E}$$

where $\pi : \mathcal{L}^{1,p}(B(x_o, \sigma R), dx) \to \mathcal{L}^{1,p}(B(x_o, \sigma R), dx)/\mathcal{Z}$ is the standard quotient map and $\mathcal{E}$ is the extension operator given by Theorem 9.1. Φ is bounded since π is bounded and $\mathcal{E}$ is bounded by Theorem 9.1. It is also easy to see that Φ is one to one by the definition of the extension operator $\mathcal{E}$. To show that Φ is onto, given an element $[f] \in \mathcal{L}^{1,p}(B(x_o, \sigma R), dx)/\mathcal{Z}$, we let $\tilde{f}$ to be any element in $[f]$ restricted to F. Clearly, $\Phi(\tilde{f}) = [f]$. It now follows that $B_\beta^p(F, d\mu)$ is a Banach space since it can be identified with the space $\mathcal{L}^{1,p}(B(x_o, \sigma R), dx)/\mathcal{Z}$. $\square$

CHAPTER 12

Returning to Carnot groups

This chapter is devoted to some applications of the theory so far developed to the setting of Carnot groups. In view of the central position of these ambients in analysis and geometry, it is desirable to present the main results in this context. We notice that the statements of the relevant theorems can be greatly simplified, due to the global character of the doubling condition, see (2.17). Also, with the exception of Theorem 12.2, we have chosen not to use an abstract measure in stating the results. Instead, we have expressed them with respect to the perimeter measure (Definition 4.7), since the latter plays a central role in the applications. In the sequel we will adopt the notations introduced in Chapter 2. We begin with a version of the interior trace inequality in Theorems 8.6, 8.8.

THEOREM 12.1. *Let $\boldsymbol{G}$ be a Carnot group. Consider of a bounded $C^{1,1}$ domain $\Omega \subset \boldsymbol{G}$ of type ≤ 2, with its perimeter measure μ. Given $p > 1$, $B_o = B(g_o, R) \subset \boldsymbol{G}$, $u \in \mathcal{L}^{1,p}(2B_o, dg)$, and $F \subset B_o$, where $F \subset \partial\Omega$, one has for any $0 < \beta \leq 1 - 1/p$*

(12.1) $$\|u\|_{B^p_\beta(F,d\mu)} \leq C \, \|u\|_{\mathcal{L}^{1,p}(2B_o,dg)} \, ,$$

for some $C = C(\boldsymbol{G}, p, \Omega, \beta) > 0$.

Proof. It follows immediately from Theorem 8.6. □

The result that follows specializes the extension Theorem 9.1.

THEOREM 12.2. *Let $\boldsymbol{G}$ be a Carnot group with homogeneous dimension Q. Suppose $1 \leq p < \infty$, and that $\Omega \subset \boldsymbol{G}$ be an open set. Assume that μ be a compactly supported, lower s-Ahlfors measure on $\boldsymbol{G}$ for some $0 < s < p$. When $p > Q$ we require in addition that $s \leq \frac{Q+p}{2}$. If $F = \text{supp } \mu$ is such that $|F| = 0$, and $F \subset \Omega$, then there exist $C > 0$, and a linear extension operator $\mathcal{E} : B^p_{1-\frac{s}{p}}(F, d\mu) \to \mathcal{L}^{1,p}(\Omega, dg)$, such that*

(i) $\mathcal{E}u(g) = u(g)$ *for μ a.e. $g \in F$,*   (ii) $\|\mathcal{E}u\|_{\mathcal{L}^{1,p}(\Omega,dg)} \leq C\|u\|_{B^p_{1-s/p}(F,d\mu)}.$

Furthermore, $\mathcal{E}u$ is supported in a neighborhood of F.

We next consider the interesting situation in which μ is the perimeter measure concentrated on the boundary of a $C^{1,1}$ sub-domain of Ω, see Definition 4.7.

THEOREM 12.3. *In a Carnot group of step 2, $\boldsymbol{G}$, consider two open sets $\tilde{\Omega} \subset\subset \Omega \subset \boldsymbol{G}$, with $\tilde{\Omega}$ a $C^{1,1}$ domain. Denoting by μ the perimeter measure associated with $\tilde{\Omega}$, for every $1 < p < \infty$ there exists a bounded linear mapping $\mathcal{E}$:*

$B^p_{1-\frac{1}{p}}(\partial \tilde{\Omega}, d\mu) \to \mathcal{L}^{1,p}(\Omega, dg)$, such that for any $u \in B^p_{1-\frac{1}{p}}(\partial \tilde{\Omega}, d\mu)$ one has

(i) $\mathcal{E}u(g) = u(g)$ for μ a.e. $g \in \partial \tilde{\Omega}$, (ii) $\|\mathcal{E}u\|_{\mathcal{L}^{1,p}(\Omega, dx)} \leq C \|u\|_{B^p_{1-s/p}(\partial \tilde{\Omega}, d\mu)}$,

for some $C = C(X, p, s, M, dist(\tilde{\Omega}, \partial \Omega)) > 0$. Furthermore, $\mathcal{E}u$ is supported in a neighborhood of $\partial \tilde{\Omega}$.

Proof. Thanks to Theorem 6.1, μ is a lower 1-Ahlfors measure. This highly non-trivial information allows to apply Theorem 12.2 with $s = 1$, $F = \partial \tilde{\Omega}$, and immediately reach the conclusion. We only need to observe that, when $p > Q$, then $\frac{Q+p}{2} > Q \geq 4 > s$. □

We now consider a version of Theorem 10.6.

THEOREM 12.4. *Let $\boldsymbol{G}$ be a Carnot group, and consider a $C^{1,1}$, connected, (ϵ, δ)-domain $\Omega \subset \boldsymbol{G}$ with its perimeter measure μ. If Ω is of type ≤ 2, then given $p > 1$, for every $0 < \beta \leq 1 - \frac{1}{p}$ there exist a linear operator*

$$Tr : \mathcal{L}^{1,p}(\Omega, dg) \to B^p_\beta(\partial \Omega, d\mu) ,$$

and a constant $C = C(\boldsymbol{G}, p, \beta, \epsilon, \delta, rad(\Omega), C^{1,1}$ character of $\Omega) > 0$, such that

(12.2) $\qquad \|Tr\, u\|_{B^p_\beta(\partial \Omega, d\mu)} \leq C \|u\|_{\mathcal{L}^{1,p}(\Omega, dg)} .$

Furthermore, if $u \in C^\infty(\overline{\Omega}) \cap \mathcal{L}^{1,p}(\Omega, dx)$, then $Tr\, u = u$ on $\partial \Omega$.

We close this chapter by considering a remarkable situation in which we can concretely characterize the traces.

THEOREM 12.5. *Let $\boldsymbol{G}$ be a Carnot group of step 2, and consider a $C^{1,1}$ connected, bounded open set Ω, with perimeter measure μ. Given $p > 1$, there exist two linear operators*

$$Tr : \mathcal{L}^{1,p}(\Omega, dg) \to B^p_{1-1/p}(\partial \Omega, d\mu) , \qquad \mathcal{E} : B^p_{1-1/p}(\partial \Omega, d\mu) \to \mathcal{L}^{1,p}(\Omega, dg) ,$$

such that $Tr \circ \mathcal{E}$ is the identity map from $B^p_{1-1/p}(\partial \Omega, d\mu)$ into itself.

Proof. The proof relies of course on Theorem 10.9, but in order to apply this result we also need to resort to various other deep facts. First of all, for a $C^{1,1}$ domain we know from Theorem 5.6 that the perimeter measure is an upper 1-Ahlfors measure. The assumption that $\boldsymbol{G}$ be of step 2 allows to enforce Theorem 6.1, and conclude that, in fact, μ is a 1-Ahlfors measure. Secondly, we need to know that $C^{1,1}$ domains in every Carnot group of step 2 are (ϵ, δ) domains. This is a highly non-trivial result. In conjunction with the development of a Fatou theory, the first study of large classes of domains in Carnot groups appeared in [**CG98**]. In that paper it was conjectured that in every Carnot group of step 2, $C^{1,1}$ domains are *non-tangentially accessible* (NTA) with respect to the CC distance. We recall the inclusion NTA $\subset (\epsilon, \delta)$. A partial answer to this conjecture was provided in [**CG98**], stating that in a Carnot group of step 2 any $C^{1,1}$ domain with cylindrical symmetry near its characteristic set is NTA, and therefore (ϵ, δ). For the Heisenberg group $\mathbb{H}^n$ a recent result of Capogna, Pauls and one of us [**CGP04**] allows to substitute the assumption of partial symmetry with the much weaker one that the characteristic points be strongly isolated. The full conjecture has been recently established by Monti and Morbidelli in their interesting paper [**MM04(I)**]. Thanks to these results, all the assumptions for the domain Ω and for the measure μ in Theorem 10.9 are fulfilled. We only need to observe that, since now $s = 1$, the hypothesis $s < p$ is trivially satisfied. Moreover, when $p > Q$, the homogeneous dimension of $\boldsymbol{G}$, then $(Q+p)/2 > Q \geq 4 > s$. We can thus implement Theorem 10.9 and reach the conclusion. □

CHAPTER 13

The Neumann problem

We consider a system of C^∞ vector fields in $\mathbb{R}^n$ satisfying the finite rank condition (1.4). Let $\Omega \subset \mathbb{R}^n$ be a bounded, open set, defined by $\Omega = \{x \in \mathbb{R}^n \mid \phi(x) < 0\}$, where $\phi : \mathbb{R}^n \to \mathbb{R}$ is C^2, and for some $\alpha > 0$

(13.1) $$|\nabla \phi(x)| \geq \alpha^{-1}, \qquad x \in \partial\Omega.$$

If ν denotes the outer unit normal to $\partial\Omega$, then $\nu = \nabla\phi/|\nabla\phi|$. Let μ be the perimeter measure supported on $\partial\Omega$ introduced in Definition 4.7. We recall, that $d\mu = |X\phi| \, dH_{n-1}\lfloor \partial\Omega$. Consider the Besov space $B^2_{\frac{1}{2}}(\partial\Omega, d\mu)$, and denote by $B^2_{\frac{1}{2}}(\partial\Omega, d\mu)^\star$ its dual space.

Formulation: *Consider the second order partial differential operator $\mathcal{L} = -\sum_{j=1}^m X_j^* X_j$. Suppose we are given $T \in B^2_{\frac{1}{2}}(\partial\Omega, d\mu)^\star$, satisfying the compatibility condition*

(13.2) $$<T, 1> = 0,$$

where $< \cdot, \cdot >$ represents the duality between $B^2_{\frac{1}{2}}(\partial\Omega, d\mu)$, and $B^2_{\frac{1}{2}}(\partial\Omega, d\mu)^\star$. The sub-elliptic Neumann problem for Ω and $\mathcal{L}$ consists in finding $u \in \mathcal{L}^{1,2}(\Omega, dx)$, such that

(13.3) $$\begin{cases} \mathcal{L}u = 0 & \text{in } \Omega, \\ \sum_{j=1}^m <X_j, \nu> X_j u = T & \text{on } \partial\Omega. \end{cases}$$

In order to introduce the appropriate formulation to this problem, let us observe that the second equation in (13.3) demands an accurate interpretation, since we are assigning the derivatives of u along the vector fields X_j, on the boundary of Ω. This term corresponds precisely to the co-normal derivative from classical elliptic theory, but now the operator $\mathcal{L}$ fails to be elliptic, and the presence of characteristic points on the boundary poses serious difficulties and must be taken in due consideration. At such points, in fact, the X_j's become tangential to $\partial\Omega$. The purpose of this chapter is to illustrate a basic application of Theorems 5.5 and 10.6, by establishing the existence of a unique (modulo constants) variational solution to (13.3). We emphasize that, differently from the weak formulation of the Dirichlet problem, that of the Neumann problem requires as a crucial prerequisite knowledge of the trace space for the appropriate Sobolev space $\mathcal{L}^{1,2}(\Omega, dx)$. This makes the Neumann problem a much harder question to tackle, since right from the beginning one cannot just resort to functional analytic tools, but one has to settle the delicate question of traces. Starting from the results in the present paper, it should be possible to undertake a deeper study of the Neumann problem, and obtain various sharp quantitative estimates of the variational solution, to whose existence we now turn. To introduce the variational formulation of the sub-ellitpic Neumann problem assume for a moment that the Neumann datum T is in $C^1(\partial\Omega)$. Suppose that u solves (13.3), and that $u \in C^2(\overline{\Omega})$ (we stress that such assumption is completely unrealistic since, near a characteristic point u experiences a dramatic loss of smoothness). First, the compatibility condition is clear, since the first equation in (13.3), and an integration by parts, give

$$\begin{aligned}
0 &= -\sum_{j=1}^{m} \int_{\Omega} X_j^* X_j u \, dx = \sum_{j=1}^{m} \int_{\Omega} X_j X_j u \, dx + \sum_{j=1}^{m} \int_{\Omega} (div \, X_j) \, X_j u \, dx \\
&= \sum_{j=1}^{m} \int_{\partial\Omega} <X_j, \nu> X_j u \, dH_{n-1} - \sum_{j=1}^{m} \int_{\Omega} (div \, X_j) \, X_j u \, dx \\
&\quad + \sum_{j=1}^{m} \int_{\Omega} (div \, X_j) \, X_j u \, dx \\
&= \sum_{j=1}^{m} \int_{\partial\Omega} <X_j, \nu> X_j u \, dH_{n-1} \\
&= \int_{\partial\Omega} T \, dH_{n-1} = <T, 1> .
\end{aligned}$$

Secondly, multiplying the first equation in (13.3) by $f \in C^\infty(\overline{\Omega})$, and integrating over Ω, we find

(13.4)
$$\begin{aligned}
0 &= -\sum_{j=1}^{m} \int_{\Omega} X_j^* X_j u \, f \, dx = \sum_{j=1}^{m} \int_{\Omega} X_j X_j u \, f \, dx + \sum_{j=1}^{m} \int_{\Omega} (div \, X_j) \, X_j u \, f \, dx \\
&= \sum_{j=1}^{m} \int_{\partial\Omega} <X_j, \nu> X_j u \, f \, dH_{n-1} - \sum_{j=1}^{m} \int_{\Omega} div(f X_j) \, X_j u \, dx \\
&\quad + \sum_{j=1}^{m} \int_{\Omega} (div \, X_j) \, X_j u \, f \, dx \\
&= \int_{\partial\Omega} T f \, dH_{n-1} - \int_{\Omega} <Xu, Xf> dx .
\end{aligned}$$

If we now assume that Ω be a domain for which there exists a continuous trace operator Tr, such that $Tr(f) = f$ for any $f \in C^\infty(\overline{\Omega})$, we can thus write

$$\int_{\partial\Omega} T f \, dH_{n-1} = <T, Tr(f)> .$$

Using the latter equation in (13.4), we finally obtain

(13.5) $\qquad \int_{\Omega} <Xu, Xf> dx = <T, Tr(f)>, \qquad$ for every $f \in C^\infty(\overline{\Omega})$.

The latter equation suggests what the variational formulation of the Neumann problem (13.3) should be. Let us pause a moment, however, to note a delicate issue connected with the boundary integral in the right-hand side of (13.4). Keeping in mind that eventually we want to take $T \in B^2_{\frac{1}{2}}(\partial\Omega, d\mu)^\star$, we wonder whether this is actually the case under the (unrealistic) smoothness assumption on u that led to (13.5). The boundary integral in the right-hand side of (13.4) is performed with respect to surface measure $H_{n-1} \lfloor \partial\Omega$, and this seems to contrast with the assumption that T belongs to the dual of the Besov space $B^2_{\frac{1}{2}}(\partial\Omega, d\mu)$, since the latter space is defined with respect to the perimeter measure $d\mu$. In view of the continuous inclusion $B^2_{\frac{1}{2}}(\partial\Omega, d\mu) \subset L^2(\partial\Omega, d\mu)$, we have from Schwarz inequality

(13.6)
$$\left| \int_{\partial\Omega} T f \, dH_{n-1} \right| \leq \left(\int_{\partial\Omega} |T|^2 \, |X\phi|^{-1} \, dH_{n-1} \right)^{1/2} \left(\int_{\partial\Omega} |f|^2 \, |X\phi| \, dH_{n-1} \right)^{1/2}$$
$$\leq \left(\int_{\partial\Omega} |T|^2 \, |X\phi|^{-1} \, dH_{n-1} \right)^{1/2} \|f\|_{B^2_{\frac{1}{2}}(\partial\Omega, d\mu)} \,.$$

From (13.6) it is thus clear that, in order to have $T \in B^2_{\frac{1}{2}}(\partial\Omega, d\mu)^*$, it would suffice to know that $T \in L^2(\partial\Omega, |X\phi|^{-1} \, dH_{n-1}) = L^2(\partial\Omega, d\mu)^*$. Let us observe explicitly, at this point, that the measure $|X\phi|^{-1} \, H_{n-1} \lfloor \partial\Omega$ can be quite singular on the characteristic set Σ, where $|X\phi| = 0$. However, we have

(13.7) $\qquad |T| = \left| \sum_{j=1}^m <X_j, \nu> X_j u \right| \leq \dfrac{|X\phi| \, |Xu|}{|\nabla\phi|} \leq \alpha \, |X\phi| \, |Xu|,$

where in the last inequality we have used (13.1). The presence of the factor $|X\phi|$ in the right-hand side of (13.7) is what saves the day, since one obtains

$$\int_{\partial\Omega} |T|^2 \, |X\phi|^{-1} \, dH_{n-1} \leq \alpha^2 \int_{\partial\Omega} |Xu|^2 \, |X\phi| \, dH_{n-1} = \alpha^2 \int_{\partial\Omega} |Xu|^2 \, d\mu < \infty \,,$$

thanks to the assumption $u \in C^2(\overline{\Omega})$. In conclusion, if we assume the existence of a solution smooth up to the boundary, the above considerations allow to interpret the boundary integral in the right-hand side of (13.4) as the action of T, in the duality of $B^2_{\frac{1}{2}}(\partial\Omega, d\mu)$, on the test function f. We are now ready to state the main result in this chapter. For simplicity, we will state it in the context of a Carnot group, since in this setting the statements of the relevant results are simpler. We stress however that Theorem 13.1 continues to be valid for an operator of Hörmander type. For a reason which will be immediately clear, we introduce the Sobolev space of functions with zero average in Ω

$$\tilde{\mathcal{L}}^{1,2}(\Omega, dx) \;=\; \left\{ f \in \mathcal{L}^{1,2}(\Omega, dx) \,|\, f_\Omega \;=\; \frac{1}{|\Omega|} \int_\Omega f(x) \, dx \;=\; 0 \right\}.$$

THEOREM 13.1. *Let $\boldsymbol{G}$ be a Carnot group of arbitrary step, and consider a $C^{1,1}$, connected, (ϵ, δ) domain $\Omega \subset \boldsymbol{G}$ of type ≤ 2. Given $T \in B^2_{\frac{1}{2}}(\partial\Omega, d\mu)^*$, satisfying the compatibility condition (13.2), there exists a unique function $u \in \tilde{\mathcal{L}}^{1,2}(\Omega, dg)$, such that*

$$\int_\Omega <Xu, Xf> dg \;=\; <T, Tr(f)>, \qquad \text{for every} \quad f \in \mathcal{L}^{1,2}(\Omega, dg).$$

We call such u the variational solution *to the sub-elliptic Neumann problem (13.3).*

Proof. We begin by observing that (ϵ, δ) domains constitute a subclass of that of Poincaré-Sobolev (PS-) domains studied in [**GN96**]. Therefore, thanks to Corollary 1.5, part II, in [**GN96**], for every $1 \leq p < \infty$, Ω supports the p-Poincaré inequality, that is, there exists a constant $C = C(\boldsymbol{G}, p) > 0$, such that

$$\int_\Omega |f - f_\Omega|^p \, dg \leq C \, (diam \, \Omega)^p \int_\Omega |Xf|^p \, dg \qquad \text{for any} \quad f \in \mathcal{L}^{1,p}(\Omega, dg).$$

This shows in particular that

$$\|f\|_{\tilde{\mathcal{L}}^{1,2}(\Omega, dg)} = \left(\int_\Omega |Xf|^2 \, dg\right)^{\frac{1}{2}}$$

defines a norm on $\tilde{\mathcal{L}}^{1,2}(\Omega, dg)$ which is equivalent to $\|\cdot\|_{\mathcal{L}^{1,2}(\Omega, dg)}$, i.e.,

(13.8) $$\|f\|_{\tilde{\mathcal{L}}^{1,2}(\Omega, dg)} \leq \|f\|_{\mathcal{L}^{1,2}(\Omega, dg)} \leq C \, \|f\|_{\tilde{\mathcal{L}}^{1,2}(\Omega, dg)}.$$

We observe next that the assumption that Ω be connected guarantees that $rad(\Omega) > 0$. Therefore, Theorem 10.7 (which follows from Theorem 5.3, and Theorem 10.6) implies the existence of a continuous trace operator

$$Tr \, : \, \mathcal{L}^{1,2}(\Omega, dg) \, \to \, B^2_{\frac{1}{2}}(\partial\Omega, d\mu),$$

such that $Tr(f) = f|_{\partial\Omega}$, for every $f \in C^\infty(\overline{\Omega})$. Consider the bilinear form $B : \tilde{\mathcal{L}}^{1,2}(\Omega, dg) \times \tilde{\mathcal{L}}^{1,2}(\Omega, dg) \to \mathbb{R}$, given by

$$B(f, g) = \int_\Omega <Xf, Xg> \, dg.$$

By (13.8), B is coercive on $\tilde{\mathcal{L}}^{1,2}(\Omega, dg)$, since

$$B(f, f) = \|f\|^2_{\tilde{\mathcal{L}}^{1,2}(\Omega, dg)}.$$

(Had we worked with the space $\mathcal{L}^{1,2}(\Omega, dg)$, we would have lost the coercivity of B, and this is why the space $\tilde{\mathcal{L}}^{1,2}(\Omega, dg)$ was introduced.) Moreover, B is obviously bounded, since an application of Schwarz inequality gives

$$|B(f, g)| \leq \|f\|_{\tilde{\mathcal{L}}^{1,2}(\Omega, dg)} \, \|g\|_{\tilde{\mathcal{L}}^{1,2}(\Omega, dg)}.$$

Next, given $T \in B^2_{\frac{1}{2}}(\partial\Omega, d\mu)^*$, satisfying (13.2), we consider the linear functional $\Lambda_T : \tilde{\mathcal{L}}^{1,2}(\Omega, dg) \to \mathbb{R}$, given by

$$\Lambda_T(f) \stackrel{def}{=} <T, Tr(f)> \, .$$

Using Theorem 10.7, we obtain

$$|\Lambda_T(f)| \leq \|T\| \, \|Tr(f)\|_{B^2_{\frac{1}{2}}(\partial\Omega, d\mu)}$$
$$\leq C \, \|T\| \, \|f\|_{\mathcal{L}^{1,2}(\Omega, dg)}$$
$$\leq C \, \|T\| \, \|f\|_{\tilde{\mathcal{L}}^{1,2}(\Omega, dg)},$$

the latter inequality being justified by (13.8). This shows $\Lambda_T \in \tilde{\mathcal{L}}^{1,2}(\Omega, dg)^*$. By the Lax-Milgram lemma, there exists a unique function $u \in \tilde{\mathcal{L}}^{1,2}(\Omega, dg)$ such that

(13.9) $$B(u, f) = \Lambda_T(f), \qquad \text{for all} \quad f \in \tilde{\mathcal{L}}^{1,2}(\Omega, dg).$$

Such u is the sought for variational solution to (13.3). We now notice that, thanks to the assumption $<T, 1> = 0$, the equation (13.9) continues to hold for any

$f \in \mathcal{L}^{1,2}(\Omega, dg)$. In fact, given such an f, we have $f - f_\Omega \in \tilde{\mathcal{L}}^{1,2}(\Omega, dg)$, and therefore (13.9) gives

$$B(u, f) = B(u, f - f_\Omega) = <T, Tr(f - f_\Omega)> = <T, Tr(f)> .$$

This completes the proof of the theorem. □

CHAPTER 14

The case of Lipschitz vector fields

In this chapter we briefly indicate how the results in the second part of this work can be generalized to the case of a system of Lipschitz vector fields in $\mathbb{R}^n$. The interest of such general setting stems from the following considerations: It includes on one hand the important case of C^∞ vector fields previously treated, on the other hand it also incorporates the general sub-elliptic operators studied in [**OR73**], [**FSC86**], since by the results in [**PS67**] the factorization matrix of a smooth positive semi-definite matrix has in general at most Lipschitz continuous entries. A further motivation comes from the fact that there are interesting classes of operators, such as, e.g., the Baouendi-Grushin ones [**Ba67**], [**Gru70**], which arise from systems of non-smooth vector fields. Remarkably, even in such general context, all the trace and extension results established in Chapters 7-11, can be obtained under the three basic assumptions, listed as (H.1),(H.2) and (H.3) below, plus an additional structural hypothesis on the CC balls, see (H.4). This is possible thanks to the general character of the theory developed here, as well as in our previous papers [**GN96**], [**GN98**], [**DGN98**]. One also needs the results in [**FW99**]. Let then $X = \{X_1, ..., X_m\}$ be a system of Lipschitz vector fields in $\mathbb{R}^n$. In order to define the CC distance associated with X, we assume that the system be controllable, at least locally. This is equivalent to saying that, with the notations of Chapter 1.1, for any connected, bounded open set $U \subset \mathbb{R}^n$, one has $S_U(x,y) \neq \varnothing$ for every $x, y \in U$. We can thus define the CC distance $d_U(x,y)$. In the sequel we fix a $\tilde{U}$ which is going to the the "universe" of our discussion, and for simplicity drop the reference to this set, and simply write $d(x,y)$. We denote by $B(x,r)$ the metric ball centered at x with radius r. Before proceeding we notice that the geometric hypothesis (1.4) becomes clearly meaningless in the present framework, and therefore some substitute assumptions are necessary. For interesting progress in this direction the reader should consult the recent article [**RSu01**] and the references therein. Following [**GN96**], [**GN98**], [**DGN98**], we next introduce the minimal topological and differential hypothesis which suffice to develop analysis in the metric space $(\mathbb{R}^n, d)$.

(H.1) $i : (\mathbb{R}^n, d_e) \to (\mathbb{R}^n, d)$ *is continuous, where* d_e *is the Euclidean distance in* $\mathbb{R}^n$.

(H.2) *For every bounded set* $U \subset \mathbb{R}^n$ *there exist constants* $C_1, R_o > 0$ *such that for* $x \in U$ *and* $0 < r < R_o$ *one has*

$$|B(x, 2r)| \leq C_1 |B(x, r)|.$$

(H.3) *Given U as in (H.2), there exist constants $C_2, R_o > 0$ such that for any $x \in U$, $0 < r < R_o$, and $f \in \mathcal{L}^{1,1}(B(x,r), dx)$, one has for $B = B(x,r)$*
$$\int_B |f(y) - f_B|\, dy \leq C_2\, r \int_B |Xf(y)|\, dy\ .$$

REMARK 14.1. In view of the results in [**GN96**], *(H.3)* can be replaced by the weaker hypothesis:

(H.3)' *With U, C_2, R_o as in (H.2), there exists $\delta \geq 1$ such that for any $x \in U$, $0 < r < R_o$ and $f \in \mathcal{L}^{1,1}(B(x,r), dx)$, one has*
$$\sup_{\lambda > 0} \{\lambda\, |\{y \in B\, |\, |f(y) - f_B| > \lambda\}|\} \leq C_2\, r \int_{\delta B} |Xf(y)|\, dy.$$

Assumptions (H.1)-(H.3) are the most basic ones. In additon to them, we assume that:

(H.4) *There exist functions $\Lambda(x,r)$, $C(x) > 0$, with $\sup\{C(x)\,|\, x \in U\} < \infty$, and a constant $\tilde{C} > 0$ such that, for any $x \in U$, and $0 < R \leq R_o$,*
$$\tilde{C}\Lambda(x, R) \leq |B(x, R)| \leq \tilde{C}^{-1}\Lambda(x, R)\ .$$

Moreover, the function Λ must satisfy
$$|\Lambda(x, r_1) - \Lambda(x, r_2)| \leq C(x) \frac{\Lambda(x, r)}{r} |r_1 - r_2|$$

for any $0 < r_1 < r_2 < R_o$, $x \in U$ and some $r \in (r_1, r_2)$.

We leave it to the interested reader to verify that, with the hypothesis (H.1)-(H.4) in force, all the results in Chapters 7-11 can be extended to the present setting.

Bibliography

[A71] D. Adams, *Trace of potentials arising from translation invariant operators*, Ann. Scuola Norm. Sup. Pisa Cl. Sci., **25** (1971), 203-217.

[A73] _____ , *A trace inequality for generalized potentials*, Studia Math., **48** (1973), 99-105.

[AH96] D. Adams & L. Hedberg, *Function Spaces and Potential Theory*, Springer-Verlag, 1996.

[Ad75] R. A. Adams, *Sobolev Spaces*, Pure and Applied Mathematics No. 65, Academic Press, New York, 1975.

[ABC93] M. Agranovsky, C. Berenstein & D.C. Chang, *Morera theorem for holomorphic H^p spaces in the Heisenberg group*, J. Reine Angew. Math., **443** (1993), 49-89.

[ABC94] _____ , *Ergodic and mixing properties of radial measures on the Heisenberg group*, Fourier analysis (Orono, ME, 1992), 1-15, Lecture Notes in Pure and Appl. Math., **157**, Dekker, New York, 1994.

[ABCP91] M. Agranovsky, C. Berenstein, D.C. Chang & D. Pascuas, *A Morera type theorem for L^2 functions in the Heisenberg group*, J. Anal. Math., **57** (1991), 282-296.

[ABCP94] _____ , *Injectivity of the Pompeiu transform in the Heisenberg group*, J. Anal. Math., **63** (1994), 131-173.

[AR99] M. L. Agranovsky & R. Rawat, *Injectivity sets for spherical means on the Heisenberg group*, J. Fourier Anal. Appl., **5** (1999), no. 4, 363-372.

[Am01] L. Ambrosio, *Some fine properties of sets of finite perimeter in Ahlfors regular metric measure spaces*, Adv. Math., **159** (2001), 51-67.

[Am02] _____ , *Fine properties of sets of finite perimeter in doubling metric measure spaces*, Calculus of variations, nonsmooth analysis and related topics. Set-Valued Anal., **10** (2002), no. 2-3, 111-128.

[BCX99] H. Bahouri, J.- Y. Chemin & C. - J. Xu, *Trace theorems in Sobolev spaces associated with Hörmander's vector fields*, Partial Differential Equations and Their Applications (Wuhan 1999), 1-14, World Sci. Publishing, River Edge, NJ, 1999.

[B00] Z. M. Balogh, *Size of characteristic sets and functions with prescribed gradients*, J. Reine Angew. Math. **564** (2003), 63-83.

[Ba67] Baouendi, M. S., *Sur une classe d'opérateurs elliptiques dégénérés*, Bull. Soc. Math. France, **95** (1967), 45-87.

[BG88] R. Beals & P. Greiner, *Calculus on Heisenberg manifolds*, Annals of Mathematics Studies, 119. Princeton University Press, Princeton, NJ, 1988.

[BGG00] R. Beals, B. Gaveau & P. C. Greiner, *Hamilton-Jacobi theory and the heat kernel on Heisenberg groups*, J. Math. Pures Appl., (9) **79** (2000), no. 7, 633-689.

[BGGV86] R. Beals, B. Gaveau, P. C. Greiner & J. Vauthier, *The Laguerre calculus on the Heisenberg group. II.*, Bull. Sci. Math., (2) **110** (1986), no. 3, 225-288.

[BGJS98] R. Beals, P. C. Greiner, Y. Jiang & L. Seco, *A functional calculus on the Heisenberg group and the boundary layer potential $\Box_+^{-1}$ for the $\bar{\partial}$-Neumann problem*, J. Funct. Anal., **155** (1998), no. 1, 205-228.

[Be96] A. Bellaïche & J.-J. Risler, ed., *Sub-Riemannian Geometry*, Birkhäuser, 1996.

[BP99] S. Berhanu & I. Pesenson, *The trace problem for vector fields satisfying Hormandër's condition*. Math. Zeit., **231** (1999), no.1, 103-122.

[Bi00] T. Bieske, *On ∞-harmonic functions on the Heisenberg Group*, Comm. Part. Diff. Eq., **27** (2002), no. 3-4, 727-761.

[BPr99] I. Birindelli & J. Prajapat, *Nonlinear Liouville type theorems in the Heisenberg group via the moving plane method*, Comm. Partial Diff. Eq., **24** (1999), 9-10, 1875-1890.

[BM95] M. Biroli & U. Mosco, *Sobolev inequalities on homogeneous spaces*, Pot. Anal., **4** (1995), no.4, 311-324.

[C76] A.-P. Calderón, *Inequalities for the maximal function relative to a metric*, Studia Math., **57** (1976), 3, 297-306.

[CDG94] L. Capogna, D. Danielli & N. Garofalo, *The geometric Sobolev embedding for vector fields and the isoperimetric inequality*, Comm. Anal. and Geom., **2** (1994), 201-215.

[CDG96] _____, *Capacitary estimates and the local behavior of solutions of nonlinear subelliptic equations*, Amer. J. Math., **118** (1996), no.6, 1153-1196.

[CDG97] _____, *Subelliptic mollifiers and a basic point-wise estimate of Poincaré type*, Math. Zeit., **226** (1997), 147-154.

[CG98] L. Capogna & N. Garofalo, *Boundary behavior of non-negative solutions of subelliptic equations in NTA domains for Carnot-Carathéodory metrics*, Journal of Fourier Anal. and Appl.,4-5, **4** (1998), 403-432.

[CG03] _____, *Regularity of minimizers of the calculus of variations in Carnot groups via hypoellipticity of systems of Hörmander type*, J. Eur. Math. Soc. (JEMS), **5** (2003), no. 1, 1-40.

[CG05] _____, *Ahlfors type estimates for perimeter measures in Carnot-Carathéodory spaces*, preprint, 2005.

[CGN98] L. Capogna, N. Garofalo & D. M. Nhieu, *A version of a theorem of Dahlberg for the subelliptic Dirichlet problem*, Math. Research Letters, **5** (1998), no. 4, 541-549.

[CGN02] _____, *Properties of harmonic measures in the Dirichlet problem for nilpotent Lie groups of Heisenberg type*, Amer. J. Math., **124** (2002), no. 2, 273-306.

[CGN04] L. Capogna, N. Garofalo & D. M. Nhieu, *Mutual absolute continuity of harmonic measure, perimeter measure and ordinary surface measure in the subelliptic Dirichlet problem*, preprint, 2005.

[CGP04] L. Capogna, N. Garofalo & S. Pauls, *Asymptotic behavior of the Gauss map near weakly singular characteristic points*, preprint, 2005.

[Ca09] C. Carathéodory, *Untersuchungen über die Grundlangen der Thermodynamik*, Math. Ann., **67** (1909), 355-386.

[Che99] J. Cheeger, *Differentiability of Lipschitz functions on metric measure spaces*, Geom. Funct. Anal. **9** (1999), no. 3, 428-517.

[CHMY04] J.H. Cheng, J.F. Hwang, A. Malchiodi & P.Yang, *Minimal surfaces in pseudohermitian geometry and the Bernstein problem in the Heisenberg group*, Ann. Scuola Norm. Sup. Pisa, to appear.

[Ch39] W. L. Chow, *Über Systeme von linearen partiellen Differentialgleichungen erster Ordnung*, Math. Annalen, **117** (1939), 98-105.

[Chr90] M. Christ, *A T(b) theorem with remarks on analytic capacity and the Cauchy integral*, Colloq. Math. **60/61** (1990), 2, 601-628.

[CGL93] G. Citti, N. Garofalo & E. Lanconelli, *Harnack's inequality for sum of squares of vector fields plus a potential*, Amer. J. Math., **115** (1993), 3, 699-734.

[CV35] Cohn-Vossen, S., *Existenz kürzester Wege*, Dokl. Akad. Nauk UzSSR **3** (1935), 339-342.

[CW71] R. Coifman & G. Weiss, *Analyse harmonique non-commutative sur certains espaces homogenes*, Springer-Verlag, (1971).

[CMZ96] T. Coulhon, D. Müller & J. Zienkiewicz, *About Riesz transforms on the Heisenberg groups*, Math. Ann., **305** (1996), no. 2, 369-379.

[CDKR91] M. Cowling, A. H. Dooley, A. Korányi & F. Ricci, *H-type groups and Iwasawa decompositions*, Adv. Math., **87** (1991), 1-41.

[CDKR] _____, *An approach to symmetric spaces of rank one via groups of Heisenberg type*, J. Geom. Anal., **8** (1998), no. 2, 199-237.

[CK84] M. Cowling & A. Korányi, *Harmonic analysis on Heisenberg type groups from a geometric viewpoint*, in "Lie Group Representation III", pp.60-100, Lec. Notes in Math., **1077** (1984), Springer-Verlag.

[Cy81] J. Cygan, *Subadditivity of homogeneous norms on certain nilpotent Lie groups*, Proc. Amer. Math. Soc., **83** (1981), 69-70.

[Da85] E. Damek, *Harmonic functions on semidirect extensions of type H nilpotent groups*, Trans. Amer. Math. Soc., 1, **290** (1985), 375-384.

[Da87] _____, *A Poisson kernel on Heisenberg type nilpotent groups*, Coll. Math., **53** (1987), 255-268.

[Da87] _____, *The geometry of semi-direct extensions of a Heisenberg type nilpotent group*, Coll. Math., **53** (1987), 255-268.

[D92] D. Danielli, *Representation formulas and embedding theorems for subelliptic operators*, C. R. Acad. Sci. Paris Sér. I Math., **314** (1992), no. 13, 987-990.

[D99] _____, *A Fefferman-Phong type inequality and applications to quasilinear subelliptic equations*, Potential Analysis, **11** (1999), no. 4, 387-413.

[DGN98] D. Danielli, N. Garofalo & D. M. Nhieu, *Trace inequalities for Carnot-Carathéodory spaces and applications*, Ann. Sc. Norm. Sup. Pisa, Cl. Sci. (4), 2, **27** (1998), 195-252.

[DGN01] _____, *Sub-elliptic Besov spaces and the characterization of traces on lower dimensional manifolds*, Harmonic Analysis and Boundary-value Problems, 2000 Arkansas Spring Lect. Series, Ed.: L.Capogna and L. Lanzani, Amer. Math. Soc., 2001.

[DGN02] _____, *A partial solution of the isoperimetric problem for the Heisenberg group*, preprint, 2002.

[DGN03(I)] _____, *Notions of convexity in Carnot groups*, Comm. Anal. Geom., **11** (2003), no. 2, 263-341.

[DGN03(II)] _____, *On the best possible character of the L^Q norm in some a priori estimates for non-divergence form equations in Carnot groups*, Proc. Amer. Math. Soc., **131** (2003), no. 11, 3487-3498.

[DGN04(I)] _____, *The theorem of Busemann-Feller-Alexandrov in Carnot groups*, Comm. Anal. Geom., **12** (2004), no. 4, 853-886.

[DGN04(II)] _____, *Minimal surfaces in Carnot groups*, preprint, 2004.

[DGS02] D. Danielli, N. Garofalo & S. Salsa, *Variational inequalities with lack of ellipticity. Part I: optimal interior regularity and non-degeneracy of the free boundary*, Indiana Univ. Math. J., **52** (2003), no. 2, 361-398.

[DGP05] D. Danielli, N. Garofalo & A. Petrosyan, *The obstacle problem in Carnot groups: $C^{1,\alpha}$ regularity of the free boundary*, preprint, 2005.

[DG54] E. De Giorgi, *Su una teoria generale della misura $(r-1)$-dimensionale in uno spazio a r dimensioni*, Ann. Mat. Pura Appl., **36** (1954), 191-213.

[DG55] _____, *Nuovi teoremi relativi alla misura $(r-1)$-dimensionale in uno spazio a r dimensioni*, Ric. Mat., **4** (1955), 95-113.

[DCP72] E. De Giorgi, F. Colombini & L. C. Piccinini, *Frontiere orientate di misura minima e questioni collegate*, Sc. Norm. Sup. Pisa, Cl. Scienze, Quaderni, 1972.

[De71] M. Derridj, *Un probléme aux limites pour une classe d'opérateurs du second ordre hypoelliptiques*, Ann. Inst. Fourier, Grenoble, **21**, 4 (1971), 99-148.

[De72] _____, *Sur un théorème de traces*, Ann. Inst. Fourier, Grenoble, **22**, 2 (1972), 73-83.

[E(I)94] P. Eberlein, *Geometry of 2-step nilpotent groups with a left invariant metric*, Ann. Sci. École Norm. Sup. (4) **27** (1994), no. 5, 611-660.

[E(II)94] _____, *Geometry of 2-step nilpotent groups with a left invariant metric. II*, Trans. Amer. Math. Soc., **343** (1994), no. 2, 805-828.

[E96] _____, *Geometry of nonpositively curved manifolds*, Chicago Lectures in Mathematics. University of Chicago Press, Chicago, IL, 1996.

[EG92] L. C. Evans & R. F. Gariepy, *Measure Theory and Fine Properties of Functions*, CRC press, 1992.

[FMM98] E. Fabes, O. Mendez & M. Mitrea, *Boundary layers on Sobolev-Besov spaces and Poisson's equation for the Laplacian in Lipschitz domains*, J. Funct. Anal., **159** (1998), 323-368.

[Fe69] H. Federer, *Geometric Measure Theory*, Springer, 1969.

[FP81] C. Fefferman & D. H. Phong, *Subelliptic eigenvalue problems*, Proceedings of the Conference in Harmonic Analysis in Honor of A. Zygmund, Wadsworth Math. Ser., Belmont, CA, (1981), 530-606.

[FSC86] C. Fefferman & A. Sanchez-Calle, *Fundamental solutions for second order subelliptic operators*, Ann. Math., **124** (1986), 247–272.

[F73] G. Folland, *A fundamental solution for a subelliptic operator*, Bull. Amer. Math. Soc., **79** (1973), 373-376.

[F75] _____, *Subelliptic estimates and function spaces on nilpotent Lie groups*, Ark. Math., **13** (1975), 161-207.

[F89] _____, *Harmonic analysis in phase space*, Annals of Mathematics Studies, **122**, Princeton University Press, Princeton, NJ, 1989.

[FS74] G. B. Folland & E. M. Stein, *Estimates for the $\bar{\partial}_b$ Complex and Analysis on the Heisenberg Group*, Comm. Pure Appl. Math., **27** (1974), 429-522.

[FGW94] B. Franchi, C. Gutiérrez & R. Wheeden, *Weighted Sobolev-Poincaré inequalities for Grushin type operators*, Comm. PDE, **19**, no.3-4, (1994), 523-604.

[FLW95] B. Franchi, G. Lu & R. L. Wheeden, *Representation formulas and weighted Poincaré inequalities for Hörmander vector fields*, Ann. Inst. Fourier Grenoble, **45** (1995), 577-604.

[FLW96] _____, *A relationship between Poincaré type inequalities and representation formulas in spaces of homogeneous type*, Inter. Math. Res. Not., **1** (1996), 1-14.

[FSS96] B. Franchi, R. Serapioni & F. Serra Cassano, *Meyers-Serrin type theorems and relaxation of variational integrals depending on vector fields*. Houston J. Math. **22** (1996), no. 4, 859-890.

[FSS01] _____, *Rectifiability and perimeter in the Heisenberg group*, Math. Ann., **321** (2001) 3, 479-531.

[FSS03(I)] _____, *On the structure of finite perimeter sets in step 2 Carnot groups.* J. Geom. Anal., **13** (2003), no. 3, 421-466.

[FSS03(II)] _____, *Regular hypersurfaces, intrinsic perimeter and implicit function theorem in Carnot groups*, Comm. Anal. Geom., **11** (2003), no. 5, 909-944.

[FSS04] _____, *Regular submanifolds, graphs and area formula in Heisenberg groups*, preprint, 2004.

[FW99] B. Franchi & R. Wheeden, *Some remarks about Poincaré type inequalities and representation formulas in metric spaces of homogeneous type*, J. Inequal. Appl. 3 (1999), no. 1, 65-89.

[Fr44] K.O. Friedrichs, *The identity of weak and strong extensions of differential operators*, Trans. Amer. Math. Soc., **55** (1944), 132-151.

[Ga57] E. Gagliardo, *Caratterizzazione delle tracce sulla frontiera relative ad alcune classi di funzioni in n variabili*, Rend. Sem. Mat. Padove, **27** (1957), 284-305.

[Ga58] _____, *Proprietà di alsune classi di funzioni in più variabili*, Ricerche Mat., **7** (1958), 102-137.

[Ga59] , _____, *Ulteriori proprietà di alcune classi di funzioni in più variabili*, **9** (1959), 24-51.

[G02] N. Garofalo, *Analysis and Geometry of Carnot-Carathéodory Spaces, With Applications to Pde's*, Birkhäuser, book in preparation.

[GL90] N. Garofalo & E. Lanconelli, *Frequency functions on the Heisenberg group, the uncertainty principle and unique continuation*, Ann. Inst. Fourier (Grenoble), **40** (1990), no. 2, 313-356.

[GL92] _____, *Existence and nonexistence results for semilinear equations on the Heisenberg group*, Indiana Univ. Math. J., **41** (1992), no. 1, 71-98.

[GN96] N. Garofalo & D. M. Nhieu, *Isoperimetric and Sobolev inequalities for Carnot-Carathéodory spaces and the existence of minimal surfaces*, Comm. Pure Appl. Math., **49** (1996), 1081-1144.

[GN98] _____, *Lipschitz continuity, global smooth approximations and extension theorems for Sobolev functions in Carnot-Carathéodory spaces*, J. Anal. Math., **74** (1998), 67-97.

[GP04] N. Garofalo & S. D. Pauls, *The Bernstein problem in the Heisenberg group*, preprint, 2004.

[GS90] N. Garofalo & F. Segala, *Estimates of the fundamental solution and Wiener's criterion for the heat equation on the Heisenberg group*, Indiana Univ. Math. J., **39** (1990), no. 4, 1155-1196.

[GT04] N. Garofalo & F. Tournier, *New properties of convex functions on the Heisenberg group*, Trans. Amer. Mat. Soc., to appear.

[GV00] N. Garofalo & D. Vassilev, *Regularity near the characteristic set in the non-linear Dirichlet problem and conformal geometry of sub-Laplacians on Carnot groups*, Math. Ann., **318** (2000), 453-516.

[GV01] _____, *Symmetry properties of positive entire solutions of Yamabe-type equations on groups of Heisenberg type*, Duke Math. J., **106** (2001), no.3, 411-448.

[Gav77] B. Gaveau, *Principe de moindre action, propagation de la chaleur et estimées sous elliptiques sur certains groupes nilpotents*, Acta Math., **139** (1977), no. 1-2, 95-153.

[GGV84] B. Gaveau, P. Greiner & J. Vauthier, *Polynomes harmoniques et problème de Dirichlet de la boule du groupe de Heisenberg en présence de symétrie radiale*, Bull. Sci. Math., (2) **108** (1984), no. 4, 337-354.

[GGV86] _____, *Intégrales de Fourier quadratiques et calcul symbolique exact sur le groupe d'Heisenberg*, J. Funct. Anal., **68** (1986), no. 2, 248-272.

[Ge77] D. Geller, *Fourier analysis on the Heisenberg group*, Proc. Nat. Acad. Sci. U.S.A., **74** (1977), no. 4, 1328-1331.

[Ge79] _____, *Necessary and sufficient conditions for local solvability on the Heisenberg group*, Harmonic analysis in Euclidean spaces (Proc. Sympos. Pure Math., Williams Coll., Williamstown, Mass., 1978), Part 2, pp. 219-226, Proc. Sympos. Pure Math., XXXV, Part, Amer. Math. Soc., Providence, R.I., 1979.

[Ge(I)80] _____, *Fourier analysis on the Heisenberg group. I. Schwartz space*, J. Funct. Anal., **36** (1980), no. 2, 205-254.

[Ge(II)80] _____, *Local solvability and homogeneous distributions on the Heisenberg group*, Comm. Part. Diff. Eq., **5** (1980), no. 5, 475-560.

[Ge(III)80] _____, *Some results in H^p theory for the Heisenberg group*, Duke Math. J.,**47** (1980), no. 2, 365-390.

[Ge84] _____, *Spherical harmonics, the Weyl transform and the Fourier transform on the Heisenberg group*, Canad. J. Math., **36** (1984), no. 4, 615-684.

[Ge90] _____, *Analytic pseudodifferential operators for the Heisenberg group and local solvability*, Mathematical Notes, **37**. Princeton University Press, Princeton, NJ, 1990.

[Ge98] _____, *Complex powers of convolution operators on the Heisenberg group*, Analysis, geometry, number theory: the mathematics of Leon Ehrenpreis (Philadelphia, PA, 1998), 223-242, Contemp. Math., **251**, Amer. Math. Soc., Providence, RI, 2000.

[GeS82] D. Geller & E. M. Stein, *Singular convolution operators on the Heisenberg group*, Bull. Amer. Math. Soc., **6** (1982), no. 1, 99-103.

[GeS84] _____, *Estimates for singular convolution operators on the Heisenberg group*, Math. Ann., **267** (1984), no. 1, 1-15.

[Gi84] E. Giusti, *Minimal surfaces and functions of bounded variation*, Birkhäuser, 1984.

[Gre80] P. C. Greiner, *Spherical harmonics on the Heisenberg group*, Canad. Math. Bull., **23** (1980), no. 4, 383-396.

[Gre81] , _____, *On the Laguerre calculus of left-invariant convolution (pseudodifferential) operators on the Heisenberg group*, Goulaouic-Meyer-Schwartz Seminar, 1980-1981, Exp. No. XI, 40 pp., École Polytech., Palaiseau, 1981.

[Gri91] Grigoryán, A. A., *The heat equation on noncompact Riemannian manifolds*, (Russian) Mat. Sb. **182** (1991), no. 1, 55-87; transl. in Math. USSR-Sb. **72** (1992), no. 1, 47-77.

[Gr85] P. Grisvard, *Elliptic problems in non-smooth domains*, Pitman, Boston, (1985).

[Gro96] M. Gromov, *Carnot-Carathéodory spaces seen from within*, in Sub-Riemannian Geometry, Progress in Mathematics, vol. 144, edited by André Bellaïche & Jean-Jacques Risler, Birkhäuser, 1996.

[Gro98] _____. *Metric Structures for Riemannian and Non-Riemannian Spaces*, Ed. by J. LaFontaine and P. Pansu, Birkhäuser, 1998.

[Gru70] V. V. Grushin, *On a class of hypoelliptic operators*, Math USSR Sbornik, **12** (1970), (3), 458-476.

[GM04] C. E. Gutiérrez & A. Montanari, *Maximum and comparison principles for convex functions on the Heisenberg group*, Comm. Part. Diff. Eq., **29** (2004), no. 9-10, 1305-1334.

[Ha96] P. Hajłasz, *Sobolev spaces on an arbitrary metric space*, Pot. Anal., **5** (1996), 403-415.

[HaM97] P. Hajłasz & O. Martio, *Traces of Sobolev functions on fractal type sets and characterization of extension domains*, J. Funct. Anal., **143** (1997), 221-246.

[HK98] P. Hajłasz & J. Kinnunen, *Hölder quasicontinuity of Sobolev functions on metric spaces*, Rev. Mat. Iberoamericana, **14** (1998), no. 3, 601-622.

[HoR92] I. Holopainen & S. Rickman, *Quasiregular mappings of the Heisenberg group*, Math. Ann., **294** (1992), no. 4, 625-643.

[H67] H. Hörmander, *Hypoelliptic second-order differential equations*, Acta Math., **119** (1967), 147-171.

[HM89] H. Hueber & D. Müller, *Asymptotics for some Green kernels on the Heisenberg group and the Martin boundary*, Math. Ann., **283** (1989), no. 1, 97-119.

[J81] _____, *The Dirichlet problem for the Kohn Laplacian on the Heisenberg group*, J. of Funct. Anal. **43** (1981), Part I, 97-141, Part II, 224-257.

[J86] D. Jerison, *The Poincaré inequality for vector fields satisfying Hörmander's condition*, Duke Math. J. 53(2), 1986, pp. 503-523.

[JL84] D. Jerison & J. Lee, *A subelliptic, nonlinear eigenvalue problem and scalar curvature on CR manifolds*, Contemporary Math., **27** (1984), 57-63.

[JL87] _____, *The Yamabe problem on CR manifolds*, J. Diff. Eq., **25** (1987), 167-197.

[JL88] _____, *Extremals for the Sobolev inequality on the Heisenberg group and the CR Yamabe problem*, J. Amer. Math. Soc., 1, **1** (1988), 1-13.

[JL89] _____, *Intrinsic CR normal coordinates and the CR Yamabe problem*, J. Diff. Geom., **29** (1989), no. 2, 303-343.

[JK95] D. Jerison & C. K. Kenig, *The inhomogeneous Dirichlet problem in Lipschitz domains*, J. Funct. Anal., **130** (1995), 1, 161-219.

[Jo81] P. W. Jones, *Quasiconformal mappings and extendability of functions in Sobolev spaces*, Acta Math., **147** (1981), 71-88.

[JW84] A. Jonsson & H. Wallin, *Function Spaces on Subsets of $\mathbb{R}^n$*, Harwood Academic, Reading, UK, 1984.

[K80] A. Kaplan, *Fundamental solutions for a class of hypoelliptic PDE generated by composition of quadratic forms*, Trans. Amer. Math. Soc., **258** (1980), 147-153.

[K81] _____, *Riemannian nilmanifolds attached to Clifford modules*, Geom. Dedicata, **11** (1981), 127-136.

[K83] _____, *On the geometry of groups of Heisenberg type*, Bull. London Math. Soc., **15** (1983), 35-42.

[KaR83] A. Kaplan & F. Ricci, *Harmonic ananlysis on groups of Heisenberg type*, in "Harmonic Analysis", pp.416-435, Lec. Notes in Math., **992** (1983), Springer-Verlag.

[Ko83] A. Korányi, *Geometric aspects of analysis on the Heisenberg group*, Topics in modern harmonic analysis, Vol. I, II (Turin/Milan, 1982), 209-258, Ist. Naz. Alta Mat. Francesco Severi, Rome, 1983.

[Ko85] _____, *Kelvin transform and harmonic polynomials on the Heisenberg group*, Adv.Math. **56** (1985), 28-38.

[Ko85] _____, *Geometric properties of Heisenberg-type groups*, Adv.Math. **56** (1985), 28-38.

[KoR85] A. Korányi & H. M. Reimann, *Quasiconformal mappings on the Heisenberg group*, Invent. Math., **80** (1985), no. 2, 309-338.

[KoR87] _____, *Horizontal normal vectors and conformal capacity of spherical rings in the Heisenberg group*, Bull. Sc. Math., **111** (1987), 3-21.

[KoR95] _____, *Foundations for the theory of quasiconformal mappings on the Heisenberg group*, Adv. Math.,

[KoS85] A. Korányi & N. Stanton, *Liouville-type theorems for some complex hypoelliptic operators*, J. Funct. Anal., **60** (1985), no. 3, 370-377.

[LaU68] O. A. Ladyzhenskaya & N. N. Uraltseva, *Linear and Quasilinear Elliptic Equations*, Academic Press, 1968.

[LM00] E. Lanconelli & D. Morbidelli, *On the Poincaré inequality for vector fields*, Ark. Mat. 38 (2000), no. 2, 327-342.

[LU98] E. Lanconelli & F. Uguzzoni, *Asymptotic behavior and non-existence theorems for semilinear Dirichlet problems involving critical exponent on unbounded domains of the Heisenberg group*, Boll. Un. Mat. Ital., (8) **1-B** (1998), 139-168.

[LU02] _____, *Degree theory for VMO maps on metric spaces and applications to Hörmander operators*, Ann. Sc. Norm. Super. Pisa Cl. Sci., (5), **1** (2002), no. 3, 569-601.

[LiMa72] J. L. Lions & M. Magenes, *Non-homogeneous boundary value problems and applications*, Vol. 1-2, Springer-Verlag, (1972).

[Lu94] G. Lu, *The sharp Poincaré inequality for free vector fields: an endpoint result*, Rev. Mat. Iberoamericana, **10** (1994), no. 2, 453-466.

[M03(I)] V. Magnani, *A blow-up theorem for regular hypersurfaces on nilpotent groups*, Manuscripta Math., **110** (2003), no. 1, 55-76.

[M03(II)] _____, *Characteristic points, rectifiability and perimeter measure on stratified groups*, preprint, 2003.

[M04] _____, *Blow-up and regular submanifolds in Heisenberg groups and applications*, preprint, 2004.

[MaSa78] O. Martio & J. Sarvas, *Injectivity theorems in the plane and space*, Ann. Acad. Sci. Fenn., **4** (1978-79), 383-401.

[M95] P. Mattila, *Geometry of sets and measures in Euclidean spaces : fractals and rectifiability*, Cambridge studies in advanced Mathematics, Cambridge University Press, **44** (1995).

[MaSC95] P. Maheux & L. Saloff-Coste, *Analyse sur les boules d'un opérateur sous-elliptique*, Math. Ann., **303** (1995), 713-740.

[Me93] M. Mekias, *Restriction to hypersurfaces of non-isotropic Sobolev spaces*, M.I.T. Ph.D Thesis, (1993).

[MZ77] N. G. Meyers & W. P. Ziemer, *Integral inequalities of Poincaré and Wirtinger type for BV functions*, Amer. J. Math., **99** (1977), 1345-1360.

[Mi00] M. Miranda, Jr., *Functions of bounded variation on "good" metric spaces*, J. Math. Pures Appl. (9) **82** (2003), no. 8, 975-1004.

[MiMa04] M. Mitrea & S. Mayboroda, *Layer Potentials and Boundary Value Problems for Laplacian in Lipschitz Domains with Data in Quasi-Banach Besov Spaces*, Ann. Mat. Pura Appl., to appear.

[MM02] R. Monti & D. Morbidelli, *Some trace theorems for vector fields*, Math. Zeit., **239** (2002), no. 4, 747-776.

[MM04(I)] _____ , *Regular domains in homogeneous groups*, Trans. Amer. Math. Soc., to appear.

[MM04(II)] _____ , *Non-tangentially accessible domains for vector fields*, Indiana Univ. Math. J., to appear.

[Mon02] R. Montgomery, *A Tour of Subriemannian Geometries, Their Geodesics and Applications*, Mathematical Surveys and Monographs, 91. American Mathematical Society, Providence, RI, 2002.

[Mo73] G. D. Mostow, *Strong Rigidity of Locally Symmetric Spaces*, Princeton Univ. Press, Princeton, N.J., 1973.

[Mu90] D. Müller, *A restriction theorem for the Heisenberg group*, Ann. of Math., (2) **131** (1990), no. 3, 567-587.

[MPR97] D. Müller, M. Peloso & F. Ricci, *On the solvability of homogeneous left-invariant differential operators on the Heisenberg group*, J. Funct. Anal., **148** (1997), no. 2, 368-383.

[MR90] D. Müller, & F. Ricci, *Analysis of second order differential operators on Heisenberg groups. I*, Invent. Math., **101** (1990), no. 3, 545-582.

[MR92] _____ , *Analysis of second order differential operators on Heisenberg groups. II*, J. Funct. Anal., **108** (1992), no. 2, 296-346.

[MRS95] D. Müller, F. Ricci & E. M. Stein, *Marcinkiewicz multipliers and multi-parameter structure on Heisenberg (-type) groups. I*, Invent. Math. **119** (1995), no. 2, 199-233.

[MRS96] _____ , *Marcinkiewicz multipliers and multi-parameter structure on Heisenberg (-type) groups. II*, Math. Z., **221** (1996), no. 2, 267-291.

[MS94] D. Müller, D. & E. M. Stein, *On spectral multipliers for Heisenberg and related groups*, J. Math. Pures Appl., (9) **73** (1994), no. 4, 413-440.

[MS99] _____ , L^p-*estimates for the wave equation on the Heisenberg group*, Rev. Mat. Iberoamericana, **15** (1999), no. 2, 297-334.

[NSW84] A. Nagel, E. M. Stein & S. Wainger, *Balls and metrics defined by vector fields I: basic properties*, Acta Math. **155** (1985), 103-147.

[NT01] E. K. Narayanan & S. Thangavelu, *Injectivity sets for spherical means on the Heisenberg group*, J. Math. Anal. Appl., **263** (2001), no. 2, 565-579.

[Ne67] J. Necas, *Les Méthodes Directes en Théorie des Équations Elliptiques*, Masson et Cie, 1967.

[NT97] A. Nevo & S. Thangavelu, *Pointwise ergodic theorems for radial averages on the Heisenberg group*, Adv. Math., **127** (1997), no. 2, 307-334.

[OR73] O. A. Oleinik & E. V. Radkevich, *Second order equations with non-negative characteristic form* (Mathematical Analysis 1969), Moscow: Itogi Nauki (1971) [Russian], English translation: Providence, R.I., Amer. Math. Soc. (1973).

[P82] P. Pansu, *Une inégalité isopérimétrique sur le groupe de Heisenberg*, C. R. Acad. Sci. Paris Sér. I Math., **295** (1982), no. 2, 127-130.

[Pa00] S. Pauls, *The large scale geometry of nilpotent Lie groups*, Comm. Anal. Geom., **1** (2000), no.2, 1-32.

[Pa(I)04] _____ , *A notion of rectifiability modeled on Carnot groups*, Indiana Univ. Math. J. 53 (2004), no. 1, 49-81.

[Pa(II)04] _____ , *Minimal surfaces in the Heisenberg group*, Geom. Dedicata, **104** (2004), 201-231.

[PS67] R. S. Phillips & L. Sarason, *Elliptic-parabolic equations of the second order*, J. Math. Mech., **17** (1967/8), 891–917.

[RSu01] F. Rampazzo & H. J. Sussmann, *Set-valued differentials and a nonsmooth version of Chow's Theorem*, Proceedings of the 40th IEEE Conference on Decision and Control; Orlando, Florida, Volume **3** (2001), pp. 2613-2618. (IEEE Publications, New York, 2001).

[Ra38] P. K. Rashevsky, *Any two points of a totally nonholonomic space may be connected by an admissible line*, Uch. Zap. Ped. Inst. im. Liebknechta, Ser. Phys. Math., (Russian) **2** (1938), 83-94.

[RT96] P. K. Ratnakumar & S. Thangavelu, *Analogues of Besicovitch-Wiener theorem for the Heisenberg group*, J. Fourier Anal. Appl., **2** (1996), no. 4, 407-414.

[RRT97] P. K. Ratnakumar, R. Rawat & S. Thangavelu, *A restriction theorem for the Heisenberg motion group*, Studia Math., **126** (1997), no. 1, 1-12.

[Ro91] C. Romero, *Potential theory for the Kohn Laplacian on the Heisenberg group*, Univ. of Minnesota Ph. D. Thesis (1991).

[RS76] L. P. Rothschild & E. M. Stein, *Hypoelliptic differential operators and nilpotent groups*, Acta Math. **137** (1976), 247–320.

[SaCo92] L. Saloff-Coste, *A note on Poincaré, Sobolev, and Harnack inequalities*, Internat. Math. Res. Notices, **2** (1992), 27-38.

[SC84] A. Sanchez-Calle, *Fundamental solutions and geometry of sum of squares of vector fields*, Inv. Math., **78** (1984), 143-160.

[Sch71] M. Schechter, *Principles of functional analysis*, Academic press, (1971).

[SST95] A. Sitaram, M. Sundari & S. Thangavelu, *Uncertainty principles on certain Lie groups*, Proc. Indian Acad. Sci. Math. Sci., **105** (1995), no. 2, 135-151.

[St61] E. M. Stein, *The characterization of functions arising as potentials I*, Bull. Amer. Math. Soc., **67** (1961), 102-104, *II*, ibidem, **68** (1962), 577-582.

[St70] _____, *Some problems in harmonic analysis suggested by symmetric spaces and semisimple groups*, Proc. Int. Congr. Math., Nice I, 1970, Gauthier-Villars, Paris, 1971, 173-179.

[St93] _____, *Harmonic Analysis: Real Variable Methods, Orthogonality and Oscillatory Integrals*, Princeton Univ. Press, (1993).

[Th90] S. Thangavelu, *Riesz means for the sub-Laplacian on the Heisenberg group*, Proc. Indian Acad. Sci. Math. Sci., **100** (1990), no. 2, 147-156.

[Th(I)91] _____, *Restriction theorems for the Heisenberg group*, J. Reine Angew. Math., **414** (1991), 51-65.

[Th(II)91] _____, *Spherical means on the Heisenberg group and a restriction theorem for the symplectic Fourier transform*, Rev. Mat. Iberoamericana, **7** (1991), no. 2, 135-155.

[Th(III)91] _____, *Some restriction theorems for the Heisenberg group*, Studia Math., **99** (1991), no. 1, 11-21.

[Th(IV)91] _____, *A multiplier theorem for the sub-Laplacian on the Heisenberg group*, Proc. Indian Acad. Sci. Math. Sci., **101** (1991), no. 3, 169-177.

[Th93] _____, *On Paley-Wiener theorems for the Heisenberg group*, J. Funct. Anal., **115** (1993), no. 1, 24-44.

[Th94] _____, *Spherical means and CR functions on the Heisenberg group*, J. Anal. Math., **63** (1994), 255-286.

[Th95] _____, *Mean periodic functions on phase space and the Pompeiu problem with a twist*, Ann. Inst. Fourier (Grenoble), **45** (1995), no. 4, 1007-1035.

[Th98] _____, *Harmonic analysis on the Heisenberg group*, Progress in Mathematics, **159**, Birkhäuser Boston, Inc., Boston, MA, 1998.

[Th00] _____, *Local ergodic theorems for K-spherical averages on the Heisenberg group*, Math. Z., **234** (2000), no. 2, 291-312.

[Th01] _____, *An analogue of Hardy's theorem for the Heisenberg group*, Colloq. Math., **87** (2001), no. 1, 137-145.

[Tre75] F. Trèves, *Basic linear partial differential equations*, Pure and Applied Mathematics, Vol. 62. Academic Press, New York-London, 1975.

[Tro87] G. M. Troianiello, *Elliptic Differential Equations and Obstacle Problems*, Plenum Press, 1987.

[V74] V. S. Varadarajan, *Lie Groups, Lie Algebras, and Their Representations*, Springer-Verlag, New York, Berlin, Heidelberg, Tokyo, 1974.

[Va86] N. Th. Varopoulos, *Analysis on Lie groups*, J. Funct. Anal., **76** (1986), 346-410.

[VSC92] N. Th. Varopoulos, L. Saloff-Coste & T. Coulhon, *Analysis and Geometry on Groups*, Cambridge U. Press, 1992.

[Zi89] W. P. Ziemer, *Weakly Differentiable Functions*, Springer-Verlag (1989).

Editorial Information

To be published in the *Memoirs*, a paper must be correct, new, nontrivial, and significant. Further, it must be well written and of interest to a substantial number of mathematicians. Piecemeal results, such as an inconclusive step toward an unproved major theorem or a minor variation on a known result, are in general not acceptable for publication. Papers appearing in *Memoirs* are generally at least 80 and not more than 200 published pages in length. Papers less than 80 or more than 200 published pages require the approval of the Managing Editor of the Transactions/Memoirs Editorial Board.

As of March 31, 2006, the backlog for this journal was approximately 13 volumes. This estimate is the result of dividing the number of manuscripts for this journal in the Providence office that have not yet gone to the printer on the above date by the average number of monographs per volume over the previous twelve months, reduced by the number of volumes published in four months (the time necessary for preparing a volume for the printer). (There are 6 volumes per year, each containing at least 4 numbers.)

A Consent to Publish and Copyright Agreement is required before a paper will be published in the *Memoirs*. After a paper is accepted for publication, the Providence office will send a Consent to Publish and Copyright Agreement to all authors of the paper. By submitting a paper to the *Memoirs*, authors certify that the results have not been submitted to nor are they under consideration for publication by another journal, conference proceedings, or similar publication.

Information for Authors

Memoirs are printed from camera copy fully prepared by the author. This means that the finished book will look exactly like the copy submitted.

The paper must contain a *descriptive title* and an *abstract* that summarizes the article in language suitable for workers in the general field (algebra, analysis, etc.). The *descriptive title* should be short, but informative; useless or vague phrases such as "some remarks about" or "concerning" should be avoided. The *abstract* should be at least one complete sentence, and at most 300 words. Included with the footnotes to the paper should be the 2000 *Mathematics Subject Classification* representing the primary and secondary subjects of the article. The classifications are accessible from www.ams.org/msc/. The list of classifications is also available in print starting with the 1999 annual index of *Mathematical Reviews*. The Mathematics Subject Classification footnote may be followed by a list of *key words and phrases* describing the subject matter of the article and taken from it. Journal abbreviations used in bibliographies are listed in the latest *Mathematical Reviews* annual index. The series abbreviations are also accessible from www.ams.org/publications/. To help in preparing and verifying references, the AMS offers MR Lookup, a Reference Tool for Linking, at www.ams.org/mrlookup/. When the manuscript is submitted, authors should supply the editor with electronic addresses if available. These will be printed after the postal address at the end of the article.

Electronically prepared manuscripts. The AMS encourages electronically prepared manuscripts, with a strong preference for $\mathcal{A}_{\mathcal{M}}\mathcal{S}$-LaTeX. To this end, the Society has prepared $\mathcal{A}_{\mathcal{M}}\mathcal{S}$-LaTeX author packages for each AMS publication. Author packages include instructions for preparing electronic manuscripts, the *AMS Author Handbook*, samples, and a style file that generates the particular design specifications of that publication series. Though $\mathcal{A}_{\mathcal{M}}\mathcal{S}$-LaTeX is the highly preferred format of TeX, author packages are also available in $\mathcal{A}_{\mathcal{M}}\mathcal{S}$-TeX.

Authors may retrieve an author package from e-MATH starting from www.ams.org/tex/ or via FTP to ftp.ams.org (login as anonymous, enter username as password, and type cd pub/author-info). The *AMS Author Handbook* and the *Instruction Manual* are available in PDF format following the author packages link from www.ams.org/tex/. The author package can also be obtained free of charge by sending email to tech-support@ams.org (Internet) or from the Publication Division, American Mathematical Society, 201 Charles St., Providence, RI 02904-2294, USA. When requesting an author package, please specify $\mathcal{AMS}$-LaTeX or $\mathcal{AMS}$-TeX and the publication in which your paper will appear. Please be sure to include your complete mailing address.

Sending electronic files. After acceptance, the source file(s) should be sent to the Providence office (this includes any TeX source file, any graphics files, and the DVI or PostScript file).

Before sending the source file, be sure you have proofread your paper carefully. The files you send must be the EXACT files used to generate the proof copy that was accepted for publication. For all publications, authors are required to send a printed copy of their paper, which exactly matches the copy approved for publication, along with any graphics that will appear in the paper.

TeX files may be submitted by email, FTP, or on diskette. The DVI file(s) and PostScript files should be submitted only by FTP or on diskette unless they are encoded properly to submit through email. (DVI files are binary and PostScript files tend to be very large.)

Electronically prepared manuscripts can be sent via email to pub-submit@ams.org (Internet). The subject line of the message should include the publication code to identify it as a Memoir. TeX source files, DVI files, and PostScript files can be transferred over the Internet by FTP to the Internet node e-math.ams.org (130.44.1.100).

Electronic graphics. Comprehensive instructions on preparing graphics are available at www.ams.org/jourhtml/graphics.html. A few of the major requirements are given here.

Submit files for graphics as EPS (Encapsulated PostScript) files. This includes graphics originated via a graphics application as well as scanned photographs or other computer-generated images. If this is not possible, TIFF files are acceptable as long as they can be opened in Adobe Photoshop or Illustrator. No matter what method was used to produce the graphic, it is necessary to provide a paper copy to the AMS.

Authors using graphics packages for the creation of electronic art should also avoid the use of any lines thinner than 0.5 points in width. Many graphics packages allow the user to specify a "hairline" for a very thin line. Hairlines often look acceptable when proofed on a typical laser printer. However, when produced on a high-resolution laser imagesetter, hairlines become nearly invisible and will be lost entirely in the final printing process.

Screens should be set to values between 15% and 85%. Screens which fall outside of this range are too light or too dark to print correctly. Variations of screens within a graphic should be no less than 10%.

Inquiries. Any inquiries concerning a paper that has been accepted for publication should be sent directly to the Electronic Prepress Department, American Mathematical Society, 201 Charles St., Providence, RI 02904, USA.

Editors

This journal is designed particularly for long research papers, normally at least 80 pages in length, and groups of cognate papers in pure and applied mathematics. Papers intended for publication in the *Memoirs* should be addressed to one of the following editors. In principle the Memoirs welcomes electronic submissions, and some of the editors, those whose names appear below with an asterisk (*), have indicated that they prefer them. However, editors reserve the right to request hard copies after papers have been submitted electronically. Authors are advised to make preliminary email inquiries to editors about whether they are likely to be able to handle submissions in a particular electronic form.

*Algebra to ALEXANDER KLESHCHEV, Department of Mathematics, University of Oregon, Eugene, OR 97403-1222; email: ams@noether.uoregon.edu

Algebra and its application to MINA TEICHER, Emmy Noether Research Institute for Mathematics, Bar-Ilan University, Ramat-Gan 52900, Israel; email: teicher@macs.biu.ac.il

Algebraic geometry to DAN ABRAMOVICH, Department of Mathematics, Brown University, Box 1917, Providence, RI 02912; email: amsedit@math.brown.edu

*Algebraic number theory to V. KUMAR MURTY, Department of Mathematics, University of Toronto, 100 St. George Street, Toronto, ON M5S 1A1, Canada; email: murty@math.toronto.edu

*Algebraic topology to ALEJANDRO ADEM, Department of Mathematics, University of British Columbia, Room 121, 1984 Mathematics Road, Vancouver, British Columbia, Canada V6T 1Z2; email: adem@math.ubc.ca

*Combinatorics to JOHN R. STEMBRIDGE, Department of Mathematics, University of Michigan, Ann Arbor, Michigan 48109-1109; email: FRS@umich.edu

Complex analysis and harmonic analysis to ALEXANDER NAGEL, Department of Mathematics, University of Wisconsin, 480 Lincoln Drive, Madison, WI 53706-1313; email: nagel@math.wisc.edu

*Differential geometry and global analysis to LISA C. JEFFREY, Department of Mathematics, University of Toronto, 100 St. George St., Toronto, ON Canada M5S 3G3; email: jeffrey@math.toronto.edu

Dynamical systems and ergodic theory to AMIE WILKINSON, Department of Mathematics, Northwestern University, 2033 Sheridan Road, Evanston, IL 60208-2730; email: transactions@math.northwestern.edu

*Functional analysis and operator algebras to MARIUS DADARLAT, Department of Mathematics, Purdue University, 150 N. University St., West Lafayette, IN 47907-2067; email: mdd@math.purdue.edu

*Geometric analysis to TOBIAS COLDING, Courant Institute, New York University, 251 Mercer St., New York, NY 10012; email: traneditor@cims.nyu.edu

*Geometric analysis to MLADEN BESTVINA, Department of Mathematics, University of Utah, 155 South 1400 East, JWB 233, Salt Lake City, Utah 84112-0090; email: bestvina@math.utah.edu

Harmonic analysis, representation theory, and Lie theory to ROBERT J. STANTON, Department of Mathematics, The Ohio State University, 231 West 18th Avenue, Columbus, OH 43210-1174; email: stanton@math.ohio-state.edu

*Logic to STEFFEN LEMPP, Department of Mathematics, University of Wisconsin, 480 Lincoln Drive, Madison, Wisconsin 53706-1388; email: lempp@math.wisc.edu

*Ordinary differential equations, and applied mathematics to PETER W. BATES, Department of Mathematics, Michigan State University, East Lansing, MI 48824-1027; email: bates@math.msu.edu

*Partial differential equations to GUSTAVO PONCE, Department of Mathematics, South Hall, Room 6607, University of California, Santa Barbara, CA 93106; email: ponce@math.ucsb.edu

*Probability and statistics to KRZYSZTOF BURDZY, Department of Mathematics, University of Washington, Box 354350, Seattle, Washington 98195-4350; email: burdzy@math.washington.edu

*Real analysis and partial differential equations to DANIEL TATARU, Department of Mathematics, University of California, Berkeley, Berkeley, CA 94720; email: tataru@math.berkeley.edu

All other communications to the editors should be addressed to the Managing Editor, ROBERT GURALNICK, Department of Mathematics, University of Southern California, Los Angeles, CA 90089-1113; email: guralnic@math.usc.edu.

Titles in This Series

860 **Thomas M. Fiore**, Pseudo limits, biadjoints, and pseudo algebras: Categorical foundations of conformal field theory, 2006

859 **N. Arcozzi, R. Rochberg, and E. Sawyer**, Carleson measures and interpolating sequences for Besov spaces on complex balls, 2006

858 **Enrico Valdinoci, Berardino Sciunzi, and Vasile Ovidiu Savin**, Flat level set regularity of p-Laplace phase transitions, 2006

857 **Donatella Danielli, Nicola Garofalo, and Duy-Minh Nhieu**, Non-doubling Ahlfors measures, perimeter measures, and the characterization of the trace spaces of Sobolev functions in Carnot-Carathéodory spaces, 2006

856 **Vladimir Bolotnikov and Harry Dym**, On boundary interpolation for matrix valued Schur functions, 2006

855 **Yevgenia Kashina, Yorck Sommerhäuser, and Yongchang Zhu**, On higher Frobenius-Schur indicators, 2006

854 **Noam Greenberg**, The role of true finiteness in the admissible recursively enumerable degrees, 2006

853 **Joachim Krieger**, Stability of spherically symmetric wave maps, 2006

852 **Viorel Barbu, Irena Lasiecka, and Roberto Triggiani**, Tangential boundary stabilization of Navier-Stokes equations, 2006

851 **Jie Wu**, On maps from loop suspensions to loop spaces and the shuffle relations on the Cohen groups, 2006

850 **Siegfried Echterhoff, S. Kaliszewski, John Quigg, and Iain Raeburn**, A categorical approach to imprimitivity theorems for C^*-dynamical systems, 2006

849 **Katsuhiko Kuribayashi, Mamoru Mimura, and Tetsu Nishimoto**, Twisted tensor products related to the cohomology of the classifying spaces of loop groups, 2006

848 **Bob Oliver**, Equivalences of classifying spaces completed at the prime two, 2006

847 **Eric T. Sawyer and Richard L. Wheeden**, Hölder continuity of weak solutions to subelliptic equations with rough coefficients, 2006

846 **Victor Beresnevich, Detta Dickinson, and Sanju Velani**, Measure theoretic laws for lim–sup sets, 2006

845 **Ehud Friedgut, Vojtech Rödl, Andrzej Ruciński, and Prasad V. Tetali**, A Sharp threshold for random graphs with a monochromatic triangle in every edge coloring, 2006

844 **Amadeu Delshams, Rafael de la Llave, and Tere M. Seara**, A geometric mechanism for diffusion in Hamiltonian systems overcoming the large gap problem: Heuristics and rigorous verification on a model, 2006

843 **Denis V. Osin**, Relatively hyperbolic groups: Intrinsic geometry, algebraic properties, and algorithmic problems, 2006

842 **David P. Blecher and Vrej Zarikian**, The calculus of one-sided M-ideals and multipliers in operator spaces, 2006

841 **Enrique Artal Bartolo, Pierrette Cassou-Noguès, Ignacio Luengo, and Alejandro Melle Hernández**, Quasi-ordinary power series and their zeta functions, 2005

840 **Sławomir Kołodziej**, The complex Monge-Ampère equation and pluripotential theory, 2005

839 **Mihai Ciucu**, A random tiling model for two dimensional electrostatics, 2005

838 **V. Jurdjevic**, Integrable Hamiltonian systems on complex Lie groups, 2005

837 **Joseph A. Ball and Victor Vinnikov**, Lax-Phillips scattering and conservative linear systems: A Cuntz-algebra multidimensional setting, 2005

836 **H. G. Dales and A. T.-M. Lau**, The second duals of Beurling algebras, 2005

835 **Kiyoshi Igusa**, Higher complex torsion and the framing principle, 2005

834 **Keníchi Ohshika**, Kleinian groups which are limits of geometrically finite groups, 2005

TITLES IN THIS SERIES

833 **Greg Hjorth and Alexander S. Kechris,** Rigidity theorems for actions of product groups and countable Borel equivalence relations, 2005

832 **Lee Klingler and Lawrence S. Levy,** Representation type of commutative Noetherian rings III: Global wildness and tameness, 2005

831 **K. R. Goodearl and F. Wehrung,** The complete dimension theory of partially ordered systems with equivalence and orthogonality, 2005

830 **Jason Fulman, Peter M. Neumann, and Cheryl E. Praeger,** A generating function approach to the enumeration of matrices in classical groups over finite fields, 2005

829 **S. G. Bobkov and B. Zegarlinski,** Entropy bounds and isoperimetry, 2005

828 **Joel Berman and Paweł M. Idziak,** Generative complexity in algebra, 2005

827 **Trevor A. Welsh,** Fermionic expressions for minimal model Virasoro characters, 2005

826 **Guy Métivier and Kevin Zumbrun,** Large viscous boundary layers for noncharacteristic nonlinear hyperbolic problems, 2005

825 **Yaozhong Hu,** Integral transformations and anticipative calculus for fractional Brownian motions, 2005

824 **Luen-Chau Li and Serge Parmentier,** On dynamical Poisson groupoids I, 2005

823 **Claus Mokler,** An analogue of a reductive algebraic monoid whose unit group is a Kac-Moody group, 2005

822 **Stefano Pigola, Marco Rigoli, and Alberto G. Setti,** Maximum principles on Riemannian manifolds and applications, 2005

821 **Nicole Bopp and Hubert Rubenthaler,** Local zeta functions attached to the minimal spherical series for a class of symmetric spaces, 2005

820 **Vadim A. Kaimanovich and Mikhail Lyubich,** Conformal and harmonic measures on laminations associated with rational maps, 2005

819 **F. Andreatta and E. Z. Goren,** Hilbert modular forms: Mod p and p-adic aspects, 2005

818 **Tom De Medts,** An algebraic structure for Moufang quadrangles, 2005

817 **Javier Fernández de Bobadilla,** Moduli spaces of polynomials in two variables, 2005

816 **Francis Clarke,** Necessary conditions in dynamic optimization, 2005

815 **Martin Bendersky and Donald M. Davis,** V_1-periodic homotopy groups of $SO(n)$, 2004

814 **Johannes Huebschmann,** Kähler spaces, nilpotent orbits, and singular reduction, 2004

813 **Jeff Groah and Blake Temple,** Shock-wave solutions of the Einstein equations with perfect fluid sources: Existence and consistency by a locally inertial Glimm scheme, 2004

812 **Richard D. Canary and Darryl McCullough,** Homotopy equivalences of 3-manifolds and deformation theory of Kleinian groups, 2004

811 **Ottmar Loos and Erhard Neher,** Locally finite root systems, 2004

810 **W. N. Everitt and L. Markus,** Infinite dimensional complex symplectic spaces, 2004

809 **J. T. Cox, D. A. Dawson, and A. Greven,** Mutually catalytic super branching random walks: Large finite systems and renormalization analysis, 2004

808 **Hagen Meltzer,** Exceptional vector bundles, tilting sheaves and tilting complexes for weighted projective lines, 2004

807 **Carlos A. Cabrelli, Christopher Heil, and Ursula M. Molter,** Self-similarity and multiwavelets in higher dimensions, 2004

806 **Spiros A. Argyros and Andreas Tolias,** Methods in the theory of hereditarily indecomposable Banach spaces, 2004

For a complete list of titles in this series, visit the
AMS Bookstore at **www.ams.org/bookstore/**.